T0258787

LANGUAGE LEARNING ONLINE

LANGUAGE LEARNING AND LANGUAGE TECHNOLOGY

Series Editors:
Carol A. Chapelle, *Iowa State University, Ames IA, USA*
Graham Davies, *Thames Valley University, London, UK*

LANGUAGE LEARNING ONLINE: TOWARDS BEST PRACTICE

EDITED BY

Ushi Felix
Monash University, Australia

Routledge
Taylor & Francis Group

LONDON AND NEW YORK

First Published in 2003 by Swets & Zeitlinger Publishers
This edition published 2014 by Routledge
2 Park Square, Milton Park, Abingdon, Oxon OX14 4RN
711 Third Avenue, New York, NY 10017, USA

Routledge is an imprint of the Taylor & Francis Group, an informa business

ISBN 978-9-026-51948-2 (hbk)
ISBN 90 265 1948 6
ISSN 1568-248X

for David

Contents

Foreword IX

Contributors 1

INTRODUCTION

1   An orchestrated vision of language learning online
     Uschi Felix 7

DESIGN

2   Language learning online: designing towards user acceptability
     Dominique Hémard 21

3   Optimising web course design for language learners
     Robert Godwin-Jones 43

TOOLS

4   Servers, clients, testing and teaching
     Stewart Arneil and Martin Holmes 59

5   Engaging the learner – how to author for best feedback
     Paul Bangs 81

6   MOO as a language learning tool
     Lesley Shield 97

7   Virtual worlds as arenas for language learning
     Patrik Svensson 123

PEDAGOGY

8   Pedagogy on the line: identifying and closing the missing links
     Uschi Felix 147

9   Using internet-based audio-graphic and video conferencing for
     language teaching and learning
     Regine Hampel and Eric Baber 171

10   Perspectives on offline and online training initiatives
     Graham Davies 193

Index 215

Foreword

For those of us who have been working in the field of learning technologies for sometime, it is possible, sometimes, to feel a little impatient with the rate of progress. Online technologies have improved in quality and power at a phenomenal rate; but we cannot say the same of online pedagogies. We can hardly blame the technologists for the rapid advance of innovative technologies, and the burgeoning opportunities for innovative practice they bring. But theirs is a more tractable field of discovery. Engineering problems require great ingenuity and perseverance, but they do promise a solution of some kind. What promise is there that we will ever understand how to motivate the majority of the population to understand the basic scientific ideas of our time? – or to make sense of numerical data? – or to speak a second language fluently? These are enduring problems, not just of our time, but from a time even before technology.

Learning technologies do hold the promise of solving these problems, but in comparison with the march of innovative technology, the march of innovative pedagogy is hardly keeping pace. Technological innovation is driven by many factors, but not one of them concerns a pedagogical imperative. There is no dialogue between teachers and technologists about what kind of technological innovation learners need: neither side knows how to begin that dialogue. It would be like asking Socrates what kind of page layout would work best: how could he advise on a technology he neither knows nor trusts? There are meaningful questions he could be asked about the design of print, but it would take a modern philosopher to ask them – someone who has a deep knowledge and experience of the technology and who also knows what Socrates knows. There are few people who can play that role for learning technology. We need to have teachers who understand both what the technology can do, and what learners need. They will be the ones to challenge it, and push it to fulfil its potential for education.

This book collects together experts from around the world who are doing this for language learning. They approach their work from different perspectives – design, tools and pedagogy – but they share the conviction that the technology is there to improve the quality and power of the learning experience. The authors of these chapters have captured a wealth of the kind of craft knowledge we need for this. For them, it is not enough to assert that online learning offers feedback to learners, for example; the assertion is challenged, elaborated, investigated, and demonstrated, in several chapters, and from different points of view. Their focus on the reality of learners' experience means that the authors are capable of forcing the technology to serve pedagogy, not vice versa.

By recording their experiences with online language learning, these innovators are able to define what counts as good practice in their field. Others should be able to build on this, but innovation in the classroom is an unequal struggle, which the technology does nothing to assist. This is the book the technologists should read if they want to understand how their innovations could help e-learning to achieve its promise forlanguage learning. Teachers need to be able to build on each others' work, to use and re-use, and re-model, and re-create, and share the results. There are few technological tools to help them do that, and a great many technological barriers. Meanwhile, we use the powerful medium of print to share their experience and recommendations, and tell the story of how they propose to mould the technology so that it really does improve the quality and power of online language learning.

Diana Laurillard

Contributors

Dominique Hémard is Principal Lecturer, Head of French at London Metropolitan University. His research interest and activities in CALL, spanning some 15 years, have led him to develop a cross-disciplinary expertise with an MSc and PhD in User Interface Design. He has been involved in designing an interactive language learning authoring tool and is particularly concerned with design issues, integration and evaluation. He has published a range of articles in international journals on hypermedia CALL, design guidelines, mental models, web-based CALL authoring and evaluation.

Robert Godwin-Jones is Professor of Foreign Languages and Faculty Fellow of the Instructional Development Center at Virginia Commonwealth University (Richmond). He has written extensively on language learning and technology. His principal interest is currently in the area of web development, with an emphasis on interactive web applications. He received in 1999 the Virginia Governor's Award in Technology as co-developer of 'Web Course in a Box', a course management system.

Stewart Arneil and Martin Holmes are the originators of the popular Half-Baked Software. They both work at the University of Victoria Humanities Computing and Media Centre in Canada and have given many Conference presentations to packed houses. Stewart is coordinator of research and development in the Media Centre and holds an MA in the philosophy of computationalism, and certification as an instructional designer. Martin is a programmer/consultant in the Centre and holds a BA (Hons) in English, an M.Phil, and the RSA DipTEFLA. He previously taught EFL for 15 years in several countries.

Paul Bangs has taught languages at school and higher education level for more than 25 years. More recently he was Director of South Bank University Language Centre in the UK, and is currently a free-lance consultant in the field of learning technology. He is an accredited expert for the European Commission (Information Society Programme), and is involved in a range of Quality Assurance work in the area of languages and technology. Over fifteen years he has been involved in the production and design of many multimedia language learning programs.

Lesley Shield is a Lecturer in Language Learning and Technology at the British Open University. In the last decade, she has worked in developing languages CD-ROMs, interactive, web-based language learning materials and the use of CMC – particularly voice-over-internet, audiographics and MOO – for language learning (at a distance). While she has published widely on the use of various CMC tools for language learning and teaching, her research focuses on MOO, in particular its collaborative and community-building aspects and the types of learning activity which best support language learning in such an environment.

Patrik Svensson is the director of HUMlab, a humanities IT lab at Umeå University in Sweden. HUMlab is a high-profile meeting place for the humanities, culture, art and new technology and new media. Many educational projects are run by the lab Swets.book Page 3 Wednesday, December 4, 2002 9:47 AM 4 and Dr. Svensson serves as project leader for a number of externally financed projects in the field of language learning and information technology. He has also co-started a national network for IT in academic language learning in Sweden and is contracted to write a book on IT and language learning for a Swedish publisher.

Uschi Felix is Associate Dean (IT) in the Faculty of Arts at Monash University in Melbourne. She is currently working on a 'best-practice' model for online language learning. She has contributed to the development of multimedia software and websites in several languages, published many critical papers in international journals on the use of technology in language teaching, and has been invited to give keynote addresses in Australia, Korea, Sweden, Finland and the UK. Her last two books Virtual Language Learning: Finding the Gems among the Pebbles and Beyond Babel: Language Learning On-line have been bestsellers.

Eric Baber is co-founder and Director of NetLearn Languages. Since its founding in 1998, NLL has been offering synchronous instruction in English, Spanish, French and a number of other languages. Since 1999 NLL has also been running courses leading to the Certificate in On-Line Teaching of English (COLTE) which expects to receive university accreditation in 2002. Eric regularly gives presentations, seminars and workshops on a range of issues revolving around the use of the Internet for educational purposes.

Graham Davies has been involved in Computer Assisted Language Learning (CALL) since 1976. In 1982 he wrote one of the first introductory books on computers in language learning and teaching, which was followed by numerous other printed and software publications. He was the Founder President of EUROCALL, holding the post from 1993 to 2000 and Chair of the WorldCALL 2003 Steering Committee. He is a partner in Camsoft, a CALL software development and consultancy business and has lectured and run ICT training courses for language teachers in 20 different countries.

Regine Hampel is a lecturer in German at the British Open University. She has been involved in developing language CD-ROMs and is currently working on the design and implementation of audio-graphic online tuition in language courses. As a member of the team involved in producing a

German course with online tuition, she received an Open University teaching
award for 2001–2002 for 'Integrating Internet-based real-time audio-graphic
conferencing tools into distance language learning'.

Diana Laurillard is Head of the e-Learning Strategy Unit at the UK
Government's Department for Education and Skills, and is Visiting Professor
at The British Open University. She previously held two terms of office as
Pro-Vice-Chancellor at the Open University, responsible for developing the
appropriate use of learning technologies within the full range of learning
and teaching methods in the University's courses. Her academic work spans
over twenty-five years of research, development and evaluation of interactive
multimedia materials and Internet services in education and training,
covering a wide range of discipline areas, including language teaching. Her
book Rethinking University Teaching, has been widely acclaimed, and is
still used as a set book in courses on learning technology all over the world.

INTRODUCTION

1

An orchestrated vision of language learning online

Uschi Felix, Monash University, Australia

Many metaphors have been used to describe the Internet or the World Wide Web, the most rehearsed being the *Information Superhighway*. This, however, immediately suggests a vision of their major function as providing information very fast and accessibly. While this function has an important role in education, it is only one of many elements in a broad spectrum of considerations which need to be orchestrated carefully in the quest for best practice in online language learning. Bateson's (1977) metaphor of the *orchestra* seems, indeed, to be much more fitting. An orchestra going through the process of learning a new piece with the ultimate goal of turning out a public performance is a useful analogy for the experience which a group of students and their teacher might share in an excellent online learning and teaching endeavour. A good conductor will facilitate and guide the enterprise, allowing players singly and in groups to shape the interpretation, and ultimately share ownership of the creation. Successful musicians will be intrinsically motivated, in tune with each other, prepared to work hard and open to new ways of reaching the desired goal, constantly updating their skills at both micro and meta levels. Their instruments will be of the best quality, finely tuned, highly reliable and supported by specialists, and the surrounding acoustics will be impeccable.

This scenario describes ideal conditions which we rarely find in educational settings, although the *Virtual Wedding* project (see Svensson this volume) comes very close. Most of us are still faced with a large number of constraints that include unmotivated students, institutional pressures, lack of time, malfunctioning technology, access problems and poor technical expertise (see Felix 2002a for a detailed discussion). The other significant difference between the orchestra rehearsal and online language learning is that in the former the outcome is always largely known. The exciting promise of some online learning environments is that the outcome will be shaped by the process in which the combination of students' interests, abilities, technological and linguistic expertise, group dynamics and serendipitous experiences contributes differently and often unexpectedly to what might have been set as a common goal. The differential outcomes of the *simulation globale* project (see Shield this volume) are clear proof of this assertion.

Not everyone would agree that this is a desirable situation, say in comparison with a structured task leading to mastery of something as tangible as the perfect tense.

Online environments have often been seen as being chaotic, with learning apparently out of control (Vogel 2001). This may well be true in some instances where collaborative online learning has suffered a similar fate to some ill-conceived communicative learning events, in which the process consists of no more than 'communicating' and the outcome of 'having communicated'. Goodyear (2002:70) puts this very succinctly:

> *Tasks designed with eyes too narrowly focused on well-specified learning*
> *outcomes may prove detrimental to the ongoing life and health of a networked*
> *learning community. Too heavy a preoccupation with the vivacity of a*
> *networked learning community may result in plenty of talk but all too little*
> *learning.*

Thankfully, the recent literature shows an ever-growing number of examples of sound pedagogical practice using the web and the Internet, both in dealing with structured tasks supporting cognitive learning approaches (Whistle 1999, Ganderton 1999, Labrie 2000, Pujolà 2001, Beaudoin 2001, Heift 2000) and open-ended, so-called 'ill-structured' projects supporting constructivist approaches (Nelson & Oliver 1999, Barson 1999, Shield et al 2001, Biesenbach-Lucas & Weasenforth 2002), or both (Erben 1999, Popov 2001, Chun & Plass 2000). The encouraging aspect of the new technologies is that they offer the potential to do both; to do either well, however, is enormously challenging and often time-consuming and costly. The metaphor of the orchestra still holds insofar as myriad elements need to be brought together in harmony in order to meet this challenge. In an extensive review of key elements in web-based instruction, Jung (2001) identified the following: *content expandability, content adaptability, visual layout, academic interaction, collaborative interaction, interpersonal interaction* and *learner autonomy*. All of these elements are touched upon in one way or another by the contributors to this volume.

Context and Definition

A clear definition of what might be meant by online language learning, or *best practice*, is almost impossible to provide. There are as many different approaches as there are nomenclatures and learning theories (Felix 2002a), and the task is not made easier by a lack of congruity between learning theory and teaching practice (Goodyear 2002). This in itself is no different from classroom learning, but we now have to deal with the added complication of the role of technology in the process.

Generally it can be said that there are two major forms of online learning. On the one hand, there are stand-alone online courses that strive to operate as virtual classrooms, in which the technology acts both as tutor and tool. High quality examples are still very rare in languages (see *Cyberitalian, Global English, Interdeutsch*). On the other hand, there are add-on activities to classroom teaching or distance education courses in which technology is used primarily as a tool and a communication device. Teachers are present to varying degrees in both forms, and both range currently from poor to excellent, just as classroom teaching does not hold an automatic monopoly on best practice education. In either context, we interpret *best practice* to mean *using the most appropriate tools to their best potential to achieve sound pedagogical processes and outcomes.*

The most important consideration in achieving best practice is the distinction between delivering static content and creating interactivity and connectivity (Felix 2002b). If all we offer online are course materials and activities in the form of drill and practice, then the harsh criticisms of bad quality teaching are well deserved and we will not have advanced from the mistakes we made in the language laboratory and early CD-ROM eras. Worse still, we will be engaging in poor pedagogy with the added frustration of poorer functionality (Arneil & Holmes and Bangs address this in detail in this volume). Equally, if all we offer is a simulation of what might be done in a classroom, then we may be underestimating the added pedagogical potential that some of the new technologies carry with them, such as exploiting authentic information gaps (see Shield this volume). With the wisdom of hindsight, though, the picture is changing and the best current approaches are driven by sound pedagogical considerations. In these the technology is being used as a tool with a variety of objectives, but two major trends can be identified. On the one hand, it is used to create learning environments in which an imaginative teacher can set up authentic learning tasks and collaborative projects, in which both the processes and the goals are stimulating and engaging, and which take individual student differences and interests into account. This is far removed from Noble's justified criticism of the content-driven, technology-as-tutor, computer-replacing-the-teacher paradigm of online learning (Noble 2001). On the other hand, the latest developments in ICALL (see the special issue, Vol 15 (5), of *CALL*) and more sophisticated handling of server-side programming, allow us to provide structured online learning activities, tailored to individual students' strengths and weaknesses and providing sophisticated feedback. The best versions of these demonstrate the technology's role in engaging students in both cognitive and metacognitive learning processes. Naturally, this latter approach also has elements of the technology-as-tutor paradigm. That the teacher will be replaced by these learning endeavours, however, is a ludicrous assumption, especially when we consider the crucial role the teacher continues to play in developing, monitoring and evaluating these sorts of activities (see Hémard this volume). If anything emerges clearly from the recent literature, it is that using technology in quality learning environments actually increases the need for teacher time and commitment (Nelson & Oliver 1999, Brabazon 2001, Morgan 2001, Rocklin 2001).

Rationale

The rationale for this book is to identify the crucial elements in the quest for best practice in online endeavours, and to discuss these in an accessible, constructively critical and realistic manner. The authors are international experts with an interest in the articulation between theory and practice, who have contributed significantly to the debate on how to exploit the new technologies in innovative and creative ways. They took on the task of considering state-of-the-art online language teaching from three crucial and overlapping perspectives – **design, tools** and **pedagogy** – with a view to formulating recommendations that can be implemented in the somewhat less than perfect educational environments of today. Clearly this group does not assume omniscience but draws from current best-practice examples; in some cases developments are so recent that it is too early to formulate what might be considered best practice, especially in the absence of empirical evidence of outcomes. Because of the large potential for content overlap, we have deliberately limited the number of contributors to allow for detailed treatment of the respective themes by each expert. At the same time, we hope that the fact that well-known problems are considered from different perspectives is a strength of the work.

Part One: Design

This section stresses the crucial importance of a design approach from the viewpoint of the researcher, as well as of the developer. *Dominique Hémard* examines conceptual considerations, frameworks, principles and guidelines, analysing concrete examples of the design process and resulting applications to shed light on how to arrive at a best practice resource informed by theory. His chapter focuses on the design considerations that underpin the concept, development and evaluation of an online learning resource. It highlights the relevance of a user-centred approach to the design process and how design theories might best be translated into practice. Using online authoring as the basis for a case study, he sets out to demonstrate that an online resource can meet the needs of language teachers and learners, thus helping to reach the critical level of acceptability necessary for integration, use and future development. His is an example where the author is simultaneously also researcher, teacher and evaluator, which can be massively time-consuming. However, the presence of the same person at all stages of the development is a very useful device in a cyclical design process in which evaluative comments feed back into the next iteration of the resource.

Considering expertise and time constraints (see Davies this volume), most teachers will need to rely on integrating existing resources or using relatively simple templates when designing their online curriculum. If all instructors want to do is add occasional interesting activities to a regular classroom program, there is a multitude of resources and materials available to them, certainly in the major European languages, and slowly increasing in some Asian languages (see Felix 2001). *Robert Godwin-Jones* looks at how a variety of these might be gathered together in the framework of a web course, either designed by the teacher or using course management systems such as *WebCT* or *Blackboard*. He examines organisation and navigation of the course environment, the complexities of content creation and how communication and interactivity might best be achieved. He also discusses the drawbacks of Learning Management Systems for language learning and suggests means of augmenting their capabilities to create the full range of sound language learning opportunities.

Part Two: Tools

This section begins with two chapters dealing with the all important question of how far sophisticated exercise routines and feedback structures can be provided online. The authors draw on experience gained in developing state-of-the-art tools such as

Hot Potatoes and *MALTED*. The need for high quality feedback is of course also discussed in other chapters and can almost be seen as a *leitmotif* leading us towards best practice. The section also includes the exciting, yet challenging and still little understood, environments of *MOOs* and *Virtual Worlds* as tools for immersing students in sophisticated collaborative and experiential learning tasks. While it might seem odd to group these four chapters together in one section, we believe that their unifying characteristic is the focus on the new technologies as *tools* in the language learning and teaching process, albeit used for quite distinct purposes and outcomes. This section illustrates the point made above about the potential of such tools for providing environments for both structured and 'ill-structured' tasks. It is also the largest section because it addresses the sorts of concerns that have given online learning the bad press it sometimes deserves: *poor online feedback, a focus on testing rather than learning, isolating learning environments, little or no interpersonal interaction, teacher-led environments and technology-driven pedagogy.*

Stewart Arneil and **Martin Holmes** lead off with a detailed discussion of the poorly understood difference between client-side and server-side exercise routines. Their chapter examines these quite different paradigms, explaining the advantages and drawbacks of each, and providing simple explanations of the underlying technology from which the strengths and limitations result. For example, they shed light on why client-side exercises cannot store results effectively due to security restrictions in the browser environment, and why server-side exercises typically allow for fewer exercise types and less control and flexibility over the exercise format and appearance. Their chapter will give instructors with limited technological skills enough information to enable them to evaluate the suitability of authoring and hosting tools for a range of typical pedagogical purposes. They also suggest ways in which the two types of technology can be combined effectively in a hybrid approach.

Paul Bangs considers the power of feedback in terms of techniques to support a sound pedagogical approach. He explains why current online feedback routines are so limited and outlines constructive, practical ways to design quality feedback within the constraints of online systems. He discusses different types of feedback, such as *extrinsic feedback* and *intrinsic feedback* and their effectiveness for different activities online. The chapter focuses on concrete examples of how to design good feedback procedures, including testing mechanisms, using the options or questions themselves to instigate the feedback process, individualising feedback, and using conditional branching. Bangs also discusses how such feedback mechanisms might be catered for by authoring systems. He examines the role of, and need for, authoring systems in general, analyses the types available, outlines advantages and limitations, and concludes with a brief description of the *MALTED* system which he sees as a significant, albeit not perfect, step forward in the quest for user-friendly yet flexible tools for creating pedagogically sound online materials.

The next two chapters deal with tools of a quite different nature. A MOO is essentially a *world built from words*. It differs from plain text chat in that participants can not only meet to chat synchronously in the MOO environment, but can also 'build' their own 'landscape' using words alone to construct it. As participants' skills in using descriptive language and, in some cases, the technical aspects of the MOO environment itself, develop, they can refine, redesign, enhance and expand their work. Indeed, the architecture of any MOO is likely to be in a constant state of evolution as players join and leave the community who 'live' and 'work' there (see also Schwienhorst 2002a on VR environments).

Though MOO offers learners the opportunity to interact both in real time and asynchronously with native and non-native speakers of the target language, with a few notable exceptions there has never been a great take-up of this technology in the language learning world. *Lesley Shield*, one of the pioneers in this field, argues that the reluctance of language teachers (and learners) to use MOO may lie in two major and widespread misconceptions about its nature. First, MOO is often thought to be a difficult technology, requiring the user to invest a great deal of time and effort in order to be able to function successfully within it. Secondly, some learners and teachers may consider the essentially text-based nature of MOO unappealing, even though interfaces that support a hypertextual and multimedia approach to MOO have emerged, allowing players to incorporate graphics, sounds and even streamed audio and video into the construction of their areas of a MOO.

Shield's chapter explores what we can learn from the literature in the field of MOO-based language learning in the context of these issues. Beginning with a technical definition of MOO, she goes on to describe what the 'non-technical' player needs to know to use the tool successfully. She also considers why MOO can be a useful tool in the learner's and the teacher's language learning toolkit. Drawing upon several case studies, the factors leading to success or failure, in terms of learning outcomes, in different MOO-based language learning activities are identified and discussed, with the roles of both learner and teacher in this medium being examined. Despite its apparently anarchic nature, she argues that MOO-based language learning can be an engaging and motivating experience as learners collaborate with others to construct, deconstruct and reconstruct their environment.

Graphical virtual environments, in contrast to the largely text-based MOOs, look like computer games and typically function as places for social interaction and different kinds of entertainment. Like MOOs, they are inherently spatial and users tend to see them as places. The main argument in *Patrik Svensson's* chapter is that the strong sense of place, the integration of multimodal media, and the possibility for students to build their own space in such worlds make graphical virtual environments promising arenas for language learning. Important points include the value of creating shared communal spaces and having language students express themselves visually as well as (hyper)textually. Moreover, the notion of *virtual* learning is problematised and Svensson argues that it is important to adopt a holistic approach. In his view, virtual reality is not unreal and reality is not and has never been unvirtual. He uses several projects as case studies in which advanced language students have produced graphical virtual representations instead of writing traditional 'paper' essays. In this environment, students become virtual construction workers, hypertext authors and community builders. While tangible outcomes in terms of linguistic work appear impressive (see the project website), another important achievement is the bridging of gaps between the linguistic, cultural and literary sub-disciplines of language teaching – and between the liberal arts and technology.

Part Three: Pedagogy

This section explores the essential elements in online pedagogy in the context of available tools, but with a focus on the student's learning processes and the teacher's role in using the latest technologies to their best potential. It also includes a chapter on training because this is often overlooked but is an essential element in good practice.

The first two chapters take a constructivist, sociocultural perspective. Even without the use of technology there has been a perceivable paradigm shift in pedagogy towards constructivist approaches, in the broadest sense, over the last two decades. Transferring to the learner responsibility for the process of learning as well as the outcome in real-life contexts where individual differences are an asset rather than a hindrance makes a great deal of motivational sense. The advent of networked and multimodal learning technologies made this shift all the more attractive – though not necessarily unproblematic in implementation (Levy 1997, Felix 1999) – because the new technologies offer meaningful learning activities even beyond what might be attempted in an excellent classroom. While most of the literature so far is speculative and anecdotal in terms of the sorts of learning outcomes that can be achieved in such ventures, rigorous studies are beginning to emerge, not only looking at differential outcomes (Warschauer 1996, Ortega 1997, Erben 1999), repair strategies (Schwienhorst 2002b) and interactivity (Labour 2001), but also at how constructivist principles are realised (Weasenforth et al 2002) in online settings. Positive effects of socially oriented factors in web-based instruction are also outlined by Jung (2001).

Our work is not meant to replicate these sorts of studies but to take a broader view of what might currently be missing in sound online pedagogy and to suggest ways in which this can be improved. The present author's own chapter identifies *a lasting sense of community, opportunities for creative speaking activities*, and *providing meaningful feedback* as the three most important areas currently ill provided for in the majority of online offerings. The chapter presents a review of current *Intelligent Language Tutoring* projects in the context of improved feedback structures and considers the role of graphics in personalising feedback. It also discusses how a sense of community might be created through the use of innovative collaborative projects and students' contribution to the learning resources, highlighting the important concepts of *social dilemma, learner versus instructor control, time management* and *authentic assessment* which arise in this context. The last section of the chapter is devoted to a detailed exploration of how two very recent voice applications, *Wimba* and *Traveler,* might be used to generate pedagogically sound oral activities online, especially in the context of alleviating language anxiety. The chapter advocates a holistic approach in an online environment – instead of separating CALL activities and interpersonal activities, which is the predominant mode in face-to-face teaching – we propose that providing all activities through the same medium, and involving students in the creation of some of them, might produce a more consistent climate of community than we find in traditional distance education and perhaps even in some classrooms.

Leading on from this exploration of two-voice applications, ***Regine Hampel*** and ***Eric Baber*** examine the pedagogy of Internet-based audio-graphic and video conferencing and analyse how the pedagogical theory can be translated into teaching practice. The technological potential for many-to-many communication online offered by these conferencing programs is realised best when supported by a pedagogical framework that goes beyond cognitive learning theories and builds on sociocultural and constructivist theories as well as taking into account the multimodal environment of audio-graphic/video conferencing. The chapter introduces three programs that allow for synchronous online voice communication to be used together with their tools (which include graphics, collaborative writing, text chat and in some cases video) and then goes on to discuss the way these programs have been implemented in both educational and commercial settings. In the light of different institutional needs, Hampel and Baber present different scenarios and course designs for using audio-graphic and video conferencing, all of which affect not only teaching methods and learning activities but also the role of the teacher. The chapter concludes with a discussion of opportunities and pitfalls which need to be taken into account if we want to develop sound pedagogical practice in audio-graphic conferencing for language learning and teaching.

Naturally none of what has been discussed so far can succeed in the absence of systematic training of language teachers (and to a lesser extent students). All too often, potentially successful existing ventures fail because teachers who have not been part of the development and planning lack the skills needed to bring them to fruition. This is particularly the case in some distance teaching contexts where not only are students geographically removed from each other, from their tutor and their institution, but tutors are also at a distance from each other as well as from course designers and software developers. Here, teacher training on the one hand and feedback loops between all those involved in the process on the other are absolutely paramount.

Teachers are of course not uniformly motivated to use technology in their teaching. This author has seen many highly successful CALL programs fail as soon as the initiator was replaced because the successor did not share the same enthusiasm for the project. Furthermore, when surveys are conducted, it is often surprising how low the general IT literacy base still is in the majority of teachers (see Davies this volume), and the frustration generated by administrators driving IT developments in institutions in the absence of equivalent support and training is often palpable (see Felix 2002a). It might be seen as a paradox to put together a book on state-of-the-art language teaching in this context. However, while we discuss relatively high-end use of technology, we hope that the scale on which this can be modified up or down and matched to teachers' and students' skills is obvious. For example, producing the routines suggested by Arneil & Holmes and Bangs would require very high level skills, whereas integrating the resulting ready-made resources such as *Hot Potatoes* or *MALTED* would be relatively simple by comparison. Similarly, many of the motivating aspects of MOOs and Virtual World environments discussed by Shield and Svensson can be achieved with simpler technologies such as discussion groups and email. Often excellent results can be achieved with the simplest applications and placing too much importance on sophisticated technologies can have a negative effect (Gibbs 2002). By the same token, it is important to look at the full spectrum of what can currently be achieved.

In our final chapter **Graham Davies** draws on his experience in training language teachers in ICT in sixteen different countries over a period of twenty years, especially his recent work in designing and managing a large body of online training materials at the *ICT4LT* website which is the outcome of an international project funded by the European Commission. His chapter does two important things: on the one hand he paints a picture of the general state of affairs in terms of realistic demands for training and how these might be addressed, touching on resource implications; on the other hand he discusses several large-scale training initiatives, most of which contain much of their materials online, questioning whether online training might be the answer to the general lack of resources. The important message is that even highly funded initiatives are not always successful, especially if individual teachers' needs are not being addressed systematically and foremost. The chapter shows clearly that training teachers fully online is likely to be as unsuccessful as teaching students fully online and recommends systematic tutor support as an important element in the success of such initiatives. It is hard not to agree with his conclusion that training is expensive but that neglecting training will prove even more expensive in the long term.

Concluding remarks

If we look at what unites these chapters, it is above all a palpable enthusiasm for the subject. These authors have a great deal of experience from which to draw but this also imbues their work with an equal measure of caution. They are fully aware of the importance of training and technical literacy and the need to use technology appropriately and not simply because it is available. They do not claim that online learning will ever replace face-to-face learning, and there is no suggestion that learning a language entirely online could ever be seen as ideal. What the book demonstrates, however, is that the new technologies offer potential for authentic encounters and constructivist learning well beyond even the best classroom simulations; that automated exercise and feedback routines do not have to be confined to drill-and-practice models but can be individualised and meaningful; and that if we are forced to teach fully by distance, these ventures no longer need to represent impoverished versions of live classes but can engender a strong sense of community. To achieve this we need to understand what elements constitute good design both in technical and pedagogical terms, to invest seriously in providing the best feedback possible, and to have the courage to take the risks associated with letting go of traditional learner/teacher relationships.

Future directions

Groucho Marx was right in claiming that the past is much easier to predict than the future. However, if we take into account developments over the last half-decade, we can identify several trends that look as if they might gain momentum. First, there is some indication that duplication of efforts is decreasing. The availability of collections of excellent resources such as the American Association of German Teachers' site *(AAGT)* is leading many teachers to integrate existing materials rather than create their own. It is hoped that world-wide *repositories of reusable or tradeable learning objects* (see Felix 2002a for more detail) will lead to more use of shared materials. Secondly, if we can be confident that automated online exercise routines are becoming more intelligent and pedagogically sound, we will be able to free up time to engage students in more extensive experiential activities. Thirdly, there has been much greater use of the new technologies for authentic encounters with the target language and culture. The trend towards task-based learning, engaging students in web quests and problem solving activities, and collaborative project-based ventures using synchronous and asynchronous text-based and graphical environments is perhaps gaining the greatest momentum. It is most likely that there will be increased communication between native and non-native speakers, whether for *rehearsal* purposes or as the only medium in which communication in the target language is likely to occur – depending of course on how we view this *third space* that we inhabit online. Fourthly, even though the possibility of engaging students in good quality oral activities online has only just offered itself, it is impossible to imagine that this will not become a standard feature of excellent online offerings, especially in distance education. The versatility of applications like *Wimba* which has just added a synchronous platform to what is discussed in Felix (this volume), and *Lyceum* (see Hampel & Baber this volume), together with increased ease of use of such products, will transform the experience of students learning a language at a distance. Lastly, and perhaps most importantly, the call for rigorous research into outcomes is becoming more urgent. We have invested enormous resources in terms of money, time and stress in languages in order to produce impressive learning materials online. While a small body of research (see Felix 2001) and our instincts lead us to believe that this investment is worthwhile, we need more large-scale studies to reassure ourselves and relevant funding bodies that real learning outcomes are being achieved. The search for appropriate research paradigms will continue as the technologies provide ever more sophisticated arenas for language learning. Chapelle's (1997:3) observation made half a decade ago has never been more apposite:

> *Despite the...implications that informative research on CALL must wait for a completely articulated theory of language teaching or psycholinguistic processing model of a second language, it is clear that the need exists for perspectives and research methods that can guide in the development and evaluation of CALL activities today.*

References

Barson, J. (1999). Dealing with Double Evolution: Action-Based Learning Approaches and Instrumental Technology. In R. Debski & M. Levy (Eds.), *WORLDCALL: Global Perspectives on Computer-Assisted Language Learning*, 11-32. Lisse: Swets & Zeitlinger.

Bateson, G. (1977). *Vers une écologie de l'esprit.* Paris: Seuil.

Beaudoin, M. (2001). Le Devoir conjugal: de la conceptualisation à la diffusion. *ALSIC*, 4 (1), 91-102.

Biesenbach-Lucas, S. & Weasenforth, D. (2002). Virtual Office Hours: Negotiation Strategies in Electronic Conferencing. *Computer Assisted Language Learning,* 15 (2), 147-166.

Brabazon, T (2001). Internet teaching and the administration of knowledge. http://www.firstmonday.org/issues/issue6_6/brabazon/index.html

Chapelle, C. (1997). Call in the year 2000: Still in search of research paradigms? *Language Learning & Technology,* 1 (1), 19-43.

Chun, D.M. & Plass, J.L. (2000). Networked multimedia environments for second language acquisition. In M. Warschauer & R. Kern, (Eds.), *Network-based Language Teaching: Concepts and Practice,* 151-170. Cambridge: Cambridge University Press.

Erben, T. (1999). Constructing Learning in a Virtual Immersion Bath: LOTE Teacher Education through Audiographics. In R. Debski & M. Levy (Eds.), *WORLDCALL: Global Perspectives on Computer-Assisted Language Learning,* 229-248. Lisse: Swets & Zeitlinger.

Felix, U. (1999). Web-Based Language Learning: A Window to the Authentic World. In R. Debski & M. Levy (Eds.), *WORLDCALL: Global Perspectives on Computer-Assisted Language Learning,* 85-98. Lisse: Swets & Zeitlinger.

Felix, U. (2001). *Beyond Babel: Language Learning Online.* Melbourne: Language Australia Ltd.

Felix, U. (2002a). Teaching languages online: Deconstructing the myths. Keynote paper presented at the *Setting the agenda: Languages, Linguistics and Area Studies in Higher Education Conference, Manchester.* In press Conference Proceedings, London: CILT.

Felix, U. (2002b). The web as a vehicle for constructivist approaches in language teaching, *ReCALL,* 14 (1), 2-15.

Ganderton, R. (1999). Interactivity in L2 Web-Based Reading. In R. Debski & M. Levy (Eds.), *WORLDCALL: Global Perspectives on Computer-Assisted Language Learning,* 49-66. Lisse: Swets & Zeitlinger.

Gibbs, J.L. (2002). *Loose coupling in global teams: Tracing the contours of cultural complexity.* Unpublished PhD dissertation, University of California, Irvine.

Goodyear, P. (2002). Psychological foundations for networked learning. In C. Steeples, & C. Jones (Eds.), *Networked Learning: Perspectives and Issues,* 49-75. London: Springer.

Heift, T. (2002). Learner Control and Error Correction in ICALL: Browsers, Peekers and Adamants. *CALICO,* 19 (3), 295-313.

Jung, I. (2001). Building a theoretical framework of web-based instruction in the context of distance education. *British Journal of Educational Technology,* 32 (5), 525-534.

Labour, M. (2001). Social constructivism and CALL: Evaluating some interactive features of network-based authoring tools. *ReCALL,* 13 (1), 32-47.

Labrie, G. (2000). A French Vocabulary Tutor for the Web. *CALICO,* 17 (3), 475-499.

Levy, M. (1997). Project-based learning for language teachers: Reflecting on the process. In R. Debski, J. Gassin & M. Smith (Eds.), *Language learning through social computing.* Occasional Papers Number 16, 181-199. Melbourne. ALAA and the Horwood Language Centre.

Morgan, C. (2001). Seeking Perseverance Through Closer Relations with Remote
    Students. Conference Proceedings (short papers), *ASCILITE'01: Meeting at the
    Crossroads,* Melbourne, Victoria, 125-128.

Nelson, T. & Oliver, W. (1999). Murder on the Internet. *CALICO,* 17 (1), 101-114.

Noble (2001). Personal interview on ABC Radio National broadcast. http://abc.net.au/
    rn/talks/bbing/mod/bbing_20012002_2856.ram

Ortega, L. (1997). Processes and outcomes in networked classroom interaction:
    Defining the research agenda for L2 computer-assisted classroom discussion.
    Language Learning & Technology, 1 (1), 82-93. http://llt.msu.edu/vol1num1/
    ortega/default.html

Popov, C. (2001). InterDeutsch-Going solo: First steps into virtual teaching on a zero
    budget. In U. Felix (Ed.), *Beyond Babel: Language Learning Online,* 15-28.
    Melbourne: Language Australia.

Pujolà, J-T. (2001). Did CALL feedback feed back? Researching learners' use of
    feedback. *ReCall,* 13 (1), 79-99.

Rocklin, T. (2001). Do I dare? Is it prudent? *National Teaching and Learning Forum
    Newsletter,* 10 (3), Oryx Press.

Schwienhorst, K. (2002a). The State of VR: A Meta-Analysis of Virtual Reality Tools
    in Second Language Acquisition. *Computer Assisted Language Learning,* 15 (3),
    221-239.

Schwienhorst, K. (2002b). Evaluating Tandem Language Learning in the MOO:
    Discourse Repair Strategies in a Bilingual Internet Project. *Computer Assisted
    Language Learning,* 15 (2), 135-146.

Shield, L., Weininger, M.J. & Davies, L.B. (2001). A task-based approach to using
    MOO for collaborative language learning. In K. Cameron (Ed.), *CALL and the
    Learning Community*, 391-401. Exeter: Elm Bank Publications.

Vogel, T. (2001). Learning out of control: Some thoughts on the World Wide Web in
    learning and teaching foreign languages. In A. Chambers & G. Davies (Eds.),
    ICT and Language Learning: A European Perspective, 133-145. Lisse: Swets &
    Zeitlinger.

Warschauer, M. (1996). Comparing face-to-face and electronic discussion in the
    second language classroom. *CALICO,* 13, 7-25.

Weasenforth, D. Biesenbach-Lucas, S. & Meloni, C. (2002). Realizing constructivist
    objectives through collaborative technologies: Threaded discussions. *Language
    Learning & Technology,* 6 (3), 58-86. http://llt.msu.edu/vol6num3/weasenforth/

Whistle, J. (1999). Concordancing with students using an 'off-the-Web' corpus.
    ReCALL, 11 (2), 74-80.

Websites

AATG – http://grow.aatg.org/index.html
Cyberitalian – http://www.cyberitalian.com/
Global English – http://www.globalenglish.com/
ICT4LT – http://www.ict4lt.org
Interdeutsch – http://www.interdeutsch.de/

All websites cited in this chapter were verified on 4.10.2002

DESIGN

2

Language learning online: designing towards user acceptability

Dominique Hémard, London Metropolitan University, U.K.

Introduction

It is widely acknowledged that Computer Assisted Language Learning (CALL), as a discipline with its own research culture and wide range of deliverables, has evolved greatly over recent years, helped by rapid technological advances. This is particularly the case with the development of multimedia, hypermedia, web-based CALL applications and computer-mediated communications. In research terms, it is equally rewarding to notice the emergence of new approaches, which are increasingly realising the importance and relevance of design considerations, alongside well-established language learning theories, to inform the design process with a view to improving the conceptual underpinning of design projects. Levy (1999), in particular, introduces the valuable concept of a design space, needed to clarify assumptions and intentions, which are, too often, ill defined by CALL developers. In turn, this design space, circumscribed by design approaches, goals, purpose and learning context, facilitates the conceptualisation of the design process within the broader context of the learning environment. This emphasis on the need to focus on important aspects of the design and its process is, indeed, very promising, if only as a necessary starting point. Yet, still too little of this message filters through to actual designs. A closer look at existing applications and the use being made of available authoring tools provides enough evidence to support the view that design is still too often driven by its perceived technological potential whilst being all too clearly affected by its own limitations. Moreover and controversially so, it could even be argued that if CALL design, use and interest have recently evolved, it is not so much as a result of a better theoretical understanding but more so because of the perceived attractiveness and opportunities generated by the web.

On this premise, the objective of this chapter is to further contextualise critical design considerations, focusing on a user-centred design approach and its practical implementations, thus presenting a conceptual process in an accessible and usable format. Ultimately, it intends to contribute to a best practice model for online language learning and teaching, instrumental in promoting a new generation of CALL developments. It is to be hoped that informed design practice will strengthen the existing research momentum by providing much needed and reliable data generated by clear and rigorous design goals. Practically, this means clarifying the adopted perspective on design and terminology, establishing a rationale for conceptualising design and demystifying the nature of the design process. The chapter will also consider the importance and relevance of knowing and understanding the users, how mental models can help in the process and how to identify and exploit user requirements. It will highlight the critical role of the designer's conceptual model and show how the theory can be successfully translated into practice by focusing on design decisions and design solutions informed by practical design tools such as design guidelines. Finally, it will show the relevance and usefulness of evaluation within an iterative design process. For greater clarity, practical examples will be used to illustrate and highlight the design process as well as relevant design issues. These will include, in particular, aspects of a CALL project, which produced a web-based authoring tool, carried out in the Department of Language Studies at the former London Guildhall University (LGU). Further details related to this CALL project can be found in Hémard & Cushion (2000a, 2000b, 2001, 2002).

Adopted perspective

> *The enormous interest in human factors of interactive systems arises from the complementary recognition of how poorly designed many current systems are and how genuinely developers desire to create elegant systems that serve the users effectively. (Shneiderman 1998:16)*

Shneiderman's comment is particularly relevant to our interest and experience in CALL design. As such, design issues in this chapter are approached from the interdisciplinary field of Human Computer Interaction (HCI) (see Levy 1997:72) which can be broadly described as 'a discipline concerned with the design, evaluation and implementation of interactive computing systems for human use and with the study of major phenomena surrounding them.' (ACM SIGCHI 1992:6). Its main tenets are that a system is more than the software alone, that its scope and purpose are wider than its functionality and that the larger system, including the human users and the physical, organisational and social environments, must be considered in order to make appropriate decisions (Preece et al 1994). It will also make references to two recently established disciplines, Interaction Design (Preece et al 2002) and Information Architecture (Kahn 2001, Veen 2000), specifically related to the study of interactive spaces and structures respectively, and therefore particularly relevant to online CALL design.

The main focus of attention will centre on the development of the conceptual design based on user requirements, its translation into a physical design and its subsequent evaluation. Throughout this chapter, reference to CALL will be made when design considerations are generically applicable across platforms or environments. However, special emphasis will be placed on the specificity of the interactive web-based CALL interface, such as web usability, access and deliveries.

Rationale for a conceptual approach to CALL design

A crucial, albeit often overlooked, concern when planning a design must be to create a CALL application which is usable. This involves developing a system which will be easy to learn, effective in what it claims to do and sufficiently motivating for users to work with it and accept its validity. This might look like stating the obvious, but too many existing applications, still, show all the hallmarks of the perseverance, enthusiasm (Cameron 1999) and personal design taste of their authors as well as the underlying expectation that they should be used within a set pedagogical framework, at the expense of a more rigorous and principled approach. If, indeed, standardisation and exposure make online CALL easier to understand and learn, is there evidence that it is effective and sufficiently enjoyable for learners to want to access it? Negative user feedback related to design features such as text scrolling, linearity and cumbersome interaction, might prove otherwise.

A system must also be conceived to be usable to reach a threshold of acceptability beyond which users can begin to interact productively and voluntarily instead of simply acting and reacting. Evidence based on much evaluative data collected over a protracted period spanning the last seven years (see, in particular, Hémard 1999, Hémard & Cushion 2001) does indeed suggest that multimedia and web interaction do not provide, by themselves, sustained attractiveness and user satisfaction. They must do so by design. Interestingly, if online interaction is still seemingly easier and safer to author than hypermedia applications, the open access and immediacy offered by the web make its usability potentially more problematic and volatile.

Furthermore, a CALL application must be rigorously designed and rationally processed, not only to be usable but also to produce valid and reliable evaluative data. What criteria can be used to evaluate an application, if specific usability goals have not been properly established in the first instance? What value will the data have if it cannot be assessed, measured or interpreted against pre-agreed benchmarks? It is only on this basis, that research in CALL and CALL design will seriously progress, capitalising on its traditional empiricist approach whilst breaking away from it.

Finally, a number of factors tend to support the case for authoring. First, authoring tools are ubiquitous and attractive as a realistic or even sole means of empowering the language teacher. Secondly, CALL as a discipline, must build its research profile despite its under-valued status and adverse political pressures (Davies 2001). CALL design and development, by encouraging collaboration as well as the cross-pollination of disciplines and generating important data is part of that research drive. Thirdly, professional designers do not necessarily provide a solution to existing CALL problems. The point must be made that small professional design teams are more likely to include graphic designers, web designers, interaction designers, animators rather than usability engineers and HCI experts. Indeed, customer satisfaction is too often gauged against the effective use of state-of-the-art functionality instead of being based on a more time-consuming understanding of the users and their interaction. Last but not least, knowing the users, in this case language teachers and learners, is an activity, which can be conducted relatively easily and effectively by CALL authors and their respective teaching departments because users are not only readily identified and approachable as interested parties, but are usually willing to play a part in CALL projects whose outcome directly concerns them. Nielsen et al (2001) corroborate this view by affirming that usability feedback can be effectively collected to improve a site even if resources, a lab, or money to hire consultants are not available.

This chapter will attempt to provide an additional design support in this quest. By focusing on selected design issues related to online CALL design and their logical outcome within a structured process, it will strive to demonstrate that successful design is possible and within easier reach than anticipated for the many struggling and financially impoverished (Felix 2001) CALL authors. Moreover, it hopes to provide the necessary theoretical basis supporting a rational and principled approach capable of withstanding scrutiny.

Demystifying the nature of the design process

Design processes, as described in the HCI literature, can be daunting due to the sheer range of expertise and the complexity of tasks involved as well as the resources required for it to be carried out successfully. Whilst there are no quick fixes, a design cycle can be made relatively simple especially for a small authoring team. In essence a successful design process will help to focus on three types of iterative activities: identify user needs and requirements; develop a design, or alternative designs if need be, meeting these requirements; and finally evaluating the chosen design's acceptability. Most importantly, as stressed by Preece et al (2002:64), 'it is important to think through and understand whether the conceptual model being developed is working in the way intended and to ensure that it is supporting the user's tasks'.

This last point can be better understood, at a higher level of abstraction, with Norman's framework (Norman 1988) comprising three interacting components: the designer, the user and the system. The design process is instigated by the designer who conceives a design model based on how the system should look and behave, which is then translated into a system image of how it should work. The process is successful when the user's understanding or model of how the system works matches the designer's. In an ideal situation, both models and the image of the system should map onto each other. Thus, users should be able to perform the tasks conceived by the designer through their interaction with the system image, since it should clearly convey the designer's model. In practice, the mapping process is difficult to achieve. If users cannot clearly understand what is expected of them or carry out a task in a way that is not intended by the designer, the system is used ineffectively and erroneously. For instance, an author may conceive an online interaction to be used in a particular order, perhaps expecting the learner to click on background information, followed by linguistic explanations and relevant applications. However, in reality, users might simply embark on a random exploration and might get lost as a result or, indeed, might simply trigger different types of links in the wrong order, crashing the computer or producing a frustrating error message.

At a more pragmatic level, the design process can be broadly confined to two iterative cycles. The first one consists in defining and refining a conceptual design on the basis of design considerations, which could be circumscribed to the following fundamental questions: what do I want to design and why, who for and how? On this basis, the first phase clarifies the problem space and the elaboration of a conceptual model, generating usability goals. In turn, the second phase focuses on specific design decisions and will generally concentrate on translating the conceptual design into a physical interface on the basis of design solutions stemming directly from the earlier design decisions. Finally, the new physical interface is fully evaluated by the users in an iterative process and modified accordingly. For instance, Version 1 of our web-based authoring tool was very much instrumental in developing further versions, on the basis of a series of iterations performed within such an iterative 'wheels within wheels' approach.

The problem space

A crucial part of the process involves examining what is intended to be designed, why it is being designed, and how the envisaged solution will help the targeted learners to overcome their identified difficulty. During this initial phase, the author should identify and circumscribe the problem space (Preece et al 1994, 2002), also referred to as design space (Levy 1999, Preece et al 1994). This includes looking into the role the computer will play, the nature of the learning environment (Levy 1999) as well as the overall shape the interaction will take. It also involves focusing on the usability of the new system by 'making explicit your implicit assumptions and claims' (Preece et al 2002:37) in the light of requirements made by the users and the problems that they have identified. For instance, to avoid previously identified navigational difficulties or disorientation when students are interacting on the web, could lead the designer to consider different types of interactive strategies (see for instance Mc Aleese 1999, Vora & Helander 1997) in order to remedy the problem. Would this be the solution? Could it be done? Working through all this systematically helps expose, at an early stage, the strengths and weaknesses of the system to be built, and as such, shape up the problem space, providing a clear design framework and, most importantly, facilitating the evaluation of the system to be designed by establishing a number of agreed usability goals and criteria which can be used as benchmarks to test the application once designed.

To illustrate this early stage of the conceptual phase, the following example highlights the elaboration of the problem space for the design and development of our web-based authoring system, which attempted to solve identified problems with existing authoring systems. The concept of an authoring tool emerged from the need to regenerate CALL activities and build a critical mass of tailor-made, web-based, interactive material within the Department of Language Studies. Previous attempts to introduce existing authoring tools had largely failed and the decision had therefore been taken to investigate the reasons why, as well as the possibility of designing our own authoring tool to entice a majority of colleagues, best described, back in 1998, as 'technophobic linguists' (Bickerton et al 2001:59). In addition to using valuable information from two surveys of CALL authors previously carried out (Hémard 1998a, Levy 1997), discussions with colleagues and students took place, based on their own CALL experience, to sketch out an author profile and identify user needs for both authors and students (Hémard & Cushion 2000b). In this particular case, the problem space was determined by these parameters and further delineated by a number of assumptions based on both authors and learners to be tested at a later stage in the process. The following four assumptions concerning authors which were made at the time are indicative of this approach. First it was held that the authoring tool would work better if it allowed its users to input their own material directly using their own word processed material since swapping between different environments had been perceived as a problem. Secondly, it was believed that it would work better for both authors and students if a standardised interactive interface without any design options was generated. This would strengthen the tool role of its functionality but also prevent personalised designs dominating and possibly antagonising students. Thirdly, it was thought that the authoring process would make more sense if authors could link up their exercises into a home page. Lastly, it was assumed that it would appeal to authors if the number of actions necessary from input to output was minimal, making the tool efficient yet simple and therefore worth using.

Similar assumptions were made concerning students. First it was held that students would feel confident to use and appreciate different types of interactive links, such as directional links when interacting with grammar exercises combined with navigational links provided by the web. Secondly, it was assumed that the students would be in a better position to contextualise the language if hypermedia links to related content material were provided. Finally, it was held that the new CALL interface would enable the students to make greater use of, and therefore better benefit from, existing open access sessions.

Design framework for a conceptual model

Thus, defining the problem space helps to establish a working design framework based on a number of assumptions linked to possible solutions to identified problems for a known targeted user group. As previously mentioned, these will centre on the following ones:

- Is there a specific problem with an existing interactive environment and why is it assumed there is such a problem?
- Who are the targeted users, what are their characteristics and requirements to remedy the problem or improve their interaction?
- What will it do? How will it support the users in their tasks?
- How will it be useful and helpful to the users? How easy will it be for them to put it into practice? How will they use it?

For whom? Or the importance and relevance of knowing and understanding the user

Approaches

Numerous methods and approaches exist to gather this important data, depending on the nature and size of the targeted user group, the type of information required, whether it should be more quantitative or qualitative, the degree of formality or informality requested and, of course, the available resources. These include the use of questionnaires, focus groups, interviews, discussions, surveys and user walkthroughs, all of which are well covered in the HCI literature (see Shneiderman 1998). However, some of these parameters in CALL are largely predetermined by the geographical location of the user group involved, the need for qualitative data and the chronic lack of resources, thus reducing the choice of methods at the disposal of the author. Bearing in mind that the main objective is to use a method or create a situation conducive to eliciting authentic views and genuine thoughts from users, we have often opted for informal discussions with colleagues and students as well as used an adapted version of the user walkthrough evaluation method. This method (see Hémard 1999, Hémard & Cushion 2001), also called verbal, *think aloud* protocols (Bainbridge 1990, Preece et al 2002, Shneiderman 1998) or cognitive walkthrough (Lewis & Wharton 1997), is primarily an evaluation technique designed to focus on the learnability and usability of a system. This method enables a designer, through the use of a prototype early in the development process or after implementation of a design, to be alerted to problems or misconceptions linked to the design of the system. The rule is simple inasmuch as users, who are paired for the exercise, are given an interactive task to perform and required to comment orally on their every move. These comments are recorded and subsequently analysed. Such a method is particularly appropriate for gathering uncorrupted qualitative data related to the interface and the user interaction but also learning approaches and general attitudes *vis à vis* CALL. Furthermore, it is realistic and manageable as it can be easily set up in a computer laboratory or even conducted in a scheduled CALL activity group. It offers the added advantage of being equally conducive to establishing and generating good collaborative work with the students concerned, facilitating in the process a fruitful exchange of information and a greater insight into the students' learning experience.

Mental models

When learning a new system or interacting in a new environment, users often relate to existing knowledge of similar systems, if they already exist, or previous experience to accelerate and simplify the learning process. Likewise, they may develop a view of how to use the new system and how it works and behaves by internalising new knowledge about it. This knowledge recall and development are referred to as mental models, triggering representations or creating new ones. Thus, an understanding of users' mental models can be particularly useful when developing a conceptual model intended to be as transparent and easy to understand for users as possible, in an attempt to reach the ultimate goal of matching both the designer's and the users' models into a perfect system image. For further evidence, refer to studies such as Allen (1997), Preece et al (1994, 2002) or Hémard (1998b).

   We have found it particularly helpful to apply the twin classification proposed by Preece et al (1994:134), which differentiates between structural and functional models. Accordingly, the structural model is formed on the development of an internal representation of the structure of the system, otherwise called *the model of 'how-it-works'*. Conversely, the functional model is developed on the strength of an internal representation of its operation, itself tagged *the model of 'how-to-use-it'*. Whereas the structural model stems from an essentially context-free, conceptual representation of the system's working structure, the functional model is generated by the association with previous comparable operations. Such a dual taxonomy can be useful for designers as the architecture of a system, as well as its tasks and expected interaction, can be mapped to better reflect users' knowledge.

Applying this design approach to online CALL, two interesting observations from previous and on-going research in hypermedia and web-based CALL design have had an important bearing on design considerations and solutions for the LGU authoring tool. First, users found it generally easier to elicit functional models related to the use of systems than structural models of how these systems work. For instance, when freely interacting with a web-based application, our students often conducted a random exploration during which navigational routes would seemingly be taken without obvious or clearly thought-out objectives. Whereas, in this instance, the browser functionality was understood, the students' own multimedia and Internet experience meant that they related to the interactive online environment as if it were a hyper-base, thus providing free access to an information base, rather than a hyper-document, which presented a specific structure with an underpinned learning strategy. Similarly, when eliciting mental models from authors, a majority of language-teaching colleagues could not relate to the spatial interactive dimension presented by the web. As a result, they simply did not know how to retrieve or even use an exercise, which had just been authored, using the authoring prototype. In a different instance, when sound was recently integrated into Version 2 of the LGU authoring tool, differences between structural and functional mental models as expressed by staff and students (Hémard & Cushion 2002) were striking. Indeed, functional models were clearly expressed, overwhelmingly projecting the functionality of the audio recorder partly associated with the audio language laboratory. Conversely, structural models were more affected by inherent problems related to sound, such as its asynchronous nature, the perceived complexity of organising and performing audio recordings and the resulting loss of control on the expected sound output. This data was paramount in determining the need for synchronisation between the authoring of textual material and its required audio input as well as its subsequent audio file management. Additional mental models expressed by teachers ranged from the audio booth with its negative experience of inputting sound associated with audio recordings, the artificiality and related quality of such recordings to the subsequent applications of these sound recordings including ways of using audio exercises more appropriately and in more flexible learning contexts. Such a categorisation made it easier to focus on the perceived problems as expressed by the users and to clarify what the specific requirements would be.

The second observation, based on data obtained over the last five years at London Guildhall University, is that mental models triggered by our students in relation to online CALL have evolved far more quickly than those of CALL authors, generating a greater discrepancy between student and author in terms of perception, experience and appreciation. This is largely due to the students' recent and sustained computer-based interaction with the Internet, seen as an intricate part of their learning experience, which has improved their functional knowledge and dexterity whilst boosting their confidence. As a result, students often expect or simply demand network access to complementary interactivity and are more capable to put their functional models of the web browser to good use by establishing the necessary links between content and language than previously anticipated. Conversely, authors' mental models of online CALL are still affected by well-entrenched demarcation lines between language and content, whilst being often blurred by a poor structural representation of its interactive potential.

Needs and requirements

Knowing and understanding the users represents an important step towards appreciating what their needs might be, in relation to their goals and aspirations. However, needs do not necessarily spring to mind or are not easy to enunciate especially when users are not aware of what is possible or what could help them to realise their tasks more effectively. In fact, from experience, users will more often know or at least find it easier to express a view about what they do not want rather than formulate precisely what their needs might be. Evidence from user walkthroughs showed that students' needs were often perceived through a reaction against or even their rejection of the interface they were interacting with. For example, from our own observations, students, on the whole, did not like the computer to tell them what to do or coerce them into completing a task before moving on to the next level. They criticised what they saw as gimmicks, complained when the processing speed was too slow or, above all, whenever the computer crashed. Similarly, they questioned the pedagogical validity of interactions and tried to trick the programme.

These comments stemming from a task-based interaction or a software demonstration are particularly useful insofar as they can trigger valuable reactions and attitudes, which can then be translated or interpreted into positive criticisms and requirements. Such a technique is not uncommon in the industry. For instance, G. Salomon, consultant interaction designer, explains that 'the best way to get started is to begin with the client doing a comprehensive demo of their product for us' (Preece et al 2002:32) thus highlighting what the existing problems and limitations are prior to formulating requirements.

Requirements

A requirement is a statement about an intended product that specifies what it should do or how it should perform. (Preece et al 2002:204)

Requirements help translate users' needs into early product specifications, which will generate a clearer understanding of its functionality and its behaviour. As such, they will range from general statements similar in tone or overlapping with high-level design principles (see section on design support below) to specific, task-based requirements. For instance, past student requirements, related to the previous example on students' needs at LGU, established that the new interface had to be easy to use and robust, that it should give users full control over their interaction as well as unrestricted movement and access. The interface was to be used as a complementary tool to their learning experience and provide a purposeful interaction coupled with a sound online support within such a language learning context (see Hémard 1999, Hémard & Cushion 2000a, 2000b). For greater adaptability and exploitation, these requirements were grouped under screen design, interactive potential of the interface and applicability for language learning.

Thus, each design phase of this particular CALL project was spearheaded by the establishment of such requirements based on feedback from the users. For instance, some requirements for Version 2 of our authoring tool were essentially functional, focusing on the need to author both text and sound simultaneously within the same interface and ensuring that the process would be manageable and easy to remember even for sporadic users. As a result, these user requirements highlighted the task of integrating and synchronising the authoring of textual material with its required audio input and the task of providing adequate and intuitive audio file management, enabling clear and easy access to stored sound files. At the interface level, requirements stemmed from the students' need for sound on demand and sound control, for a higher quality of sound delivery than previously experienced in language laboratory conditions, for an easy to use system compared to the cumbersome and inaccurate functionality of the language laboratory tape recorder and for quick and easy access via the network for private practice if required.

Summary of techniques

The following table represents an experiential appreciation of the techniques, which have been used in our CALL design projects to elicit mental models and identify needs:

Table 1. Techniques used to elicit mental models and needs

Techniques	*Nature of Data*	*Advantages*	*Disadvantages*
Question-naires	Qualitative and quantitative.	Easy and expedient; generally cheap to administer; can reach a large number of users.	Data can be erroneous, impressionistic or prone to interpretations (students are notorious for not filling in questionnaires properly).
One to one discussions	Essentially qualitative	Focused and targeted; facilitate exchange thus can glean authentic, thus valuable data.	Very time consuming; difficult to entice users to set time aside; provides specific data, which might not be representative.

Techniques	Nature of Data	Advantages	Disadvantages
User walk-throughs	Essentially qualitative, though can be quantitative (number of keystrokes, tracking through a task)	Closest to an authentic and natural setting, thus eliminates apprehension. Generate genuine and reliable data, based on direct interaction and immediate reactions. Facilitate co-operation and exchange. Produce large amounts of data. Can be fun and informal.	Data analysis is time consuming. Can lead to a resource issue if users are remunerated for their participation. May require technical support, especially if a video camera is used to tape the interactive progress made on screen.
Focus groups	Essentially qualitative	Can stimulate fruitful discussion and highlight agreements and disagreements.	Very difficult to arrange in practice with a volatile student population or busy colleagues. Requires special skills to run adequately. Too formal.
Observations	Qualitative	Easy to do; capitalise on the dual teacher / author position; give an important and unique insight into the context of use.	Not always easy to use or interpret the data; can lead to subjective or parochial analyses; time consuming.
Evaluation of existing CALL software	Qualitative and quantitative	Provides added insight into existing good practice as well as design problems; helps develop a principled and structured approach to design.	Time consuming; requires access to software, which can be expensive.

From requirements to usability goals

Requirements, once identified and selected, become useful usability goals, making the transfer from theory to practice far easier as these are subsequently translated into design decisions and solutions and present concrete goals against which to evaluate the newly designed software. For example, these goals, also called usability criteria or metrics when measurable and specific (Preece et al 1994), can include a specific completion time or a finite number of keystrokes to realise a task. In so doing, the gathering of appropriate requirements is crucial to the evaluation process, which will attempt to assess the support provided and measure performance against those established indicators.

For example, earlier requirements related to the CALL authoring project at LGU were subsequently translated into usability goals such as: Teachers with little ICT experience must always be in a position to input their language material from their known word processing environment into the authoring platform; the functionality must be justified at all times; provide a mechanism such as the creation of a Home page to facilitate the hypertext linking of authored interactive exercises; no more than ten keystrokes would be necessary to complete one itemised authoring cycle from clicking on the application's icon on the desktop to saving the authored item as a new file.

Towards designing a conceptual model

Good user interfaces start with clean, simple, task-oriented conceptual models. The conceptual model is the bones of the design. (Johnson & Henderson 2000)

The conceptual model is a high level description of what the system to be designed is supposed to do and how it should behave. Johnson & Henderson (2002) liken the conceptual model to an *idealised* view of how the system works with its *ontological* structure and mechanism for accomplishing the tasks the system intends to support. Thus, it will essentially focus on the concept underpinning the design as well as the necessary mapping between such a concept and the perceived tasks to be performed by users. In this respect, it is a crucial and central part of the design process as it provides the basis for early usability testing and, furthermore, will help shape design decisions necessary for the design and implementation of the user interface. Above all, it is an important phase of the first cycle of the design process as the conceptual model will help ensure, through early iterative testing, that the designed product is understandable by users in the way it was intended to.

What will it do? Structure and tasks

Once satisfied that usability requirements, goals and criteria have been adequately identified, it becomes possible to describe the interactive context within which the system will be used, consider the type of architecture the system should adopt, and specify what tasks it will achieve and how. Metaphors can also be called upon to facilitate the conceptualisation of the system depending on the required activities.

Many techniques exist to clarify and present the conceptual behaviour of the perceived application. They include the use of scenarios and use cases in which the various activities or tasks are described in a story form or through the interactive path between the user and the system. A third and preferred method consists of resorting to visual charts and diagrams such as the *Structure Chart* and the *Hierarchical Task Analysis Diagram* (see Preece et al 1994), which, in our own experience, are easy to produce, require minimal expertise and have the definite advantage of creating a greater visual impact. Such a visual support also helps the designer to focus on the structural map of the system, facilitating the presentation and justification of its overall strategy, as well as its more functional task-based role. Such a task analysis will also be instrumental in creating an early prototype necessary for the formative evaluation with users and experts alike.

Is there a suitable metaphor to help support the interface?

The web is not like TV. Most fundamentally, the web is a user-driven, narrow-casting medium utilising low band-width with high flexibility. (Nielsen 2000:365)

Metaphors and analogies are ready made conceptual models, to which users can easily relate and which can be called upon when conceptualising the interface. Indeed, they can be useful for three reasons. First, they can facilitate the understanding of a new or difficult concept by introducing a familiar visualisation with similar behaviours and properties (Preece et al 2002). Secondly, resorting to a metaphor can give the interactive environment to be designed a greater sense of identity and a more unifying framework (Nielsen 2000). Finally, thinking in terms of metaphors may help clarify the type of interface or interaction which might be more appropriate for the identified problem space (Erickson 1990). In this respect, metaphors could be conveniently and usefully grouped into different taxonomies depending on the nature of their conceptualisation. The first type could conceptualise a style of interaction and include the broader metaphor of the computer performing as a tool, as a tutor, or as a game. A second type could refer to an aspect of the interface such as the desktop, the browser but also the window, the frame or scrolling. Consider for instance the strong physical identification with the book, the card or the flowchart metaphors of existing authoring platforms, which are designed to help the user relate to the chosen concept of authoring but which are similarly dangerous due to their overwhelming design influence. A third type of metaphors could be used to visualise the role of the computer in a wider setting such as the Internet highway, the portal or the gateway, emphasising navigation and exploration. Others would provide mental simplifications of operations like cut, copy and paste or resort to iconic representations such as the hand, the bin but also the tree or steps to represent navigational modes or even symbols for cultural representations.

Such taxonomies can be useful inasmuch as they can trigger a necessary conceptual visualisation process in the author. What should the interface look like? Will it have navigational and content frames or menus? Will it promote hierarchical links, a scaffolding approach or exploratory navigation? Will it accommodate the use of icons or not? However, metaphors can also be misleading and prevent users from fully understanding a system's functionality. Furthermore, they can convey an overpowering model to users and, as such, limit their interaction as well as the imagination of the author.

The critical role of the designer's conceptual model

By the end of this first cycle, the designer will have identified the target user group or groups, understood what their relevant experience, impressions, apprehensions and claims were, circumscribed the problem space based on its new usability remit and behavioural limits, carried out a formative evaluation of this early conceptual prototype, and clarified and agreed on the important usability goals and criteria defining this problem space. This process being iterative, this first conceptual blue print will probably have had to have gone back to the drawing board if the outcome of the formative evaluation was inconclusive or, indeed, if the match between user requirements and realisation proved impossible or unacceptable. Crucially, a conceptual model, once tested and agreed, will not only pave the way for the physical design, providing a theoretical basis for the application of design supports, such as design principles and guidelines, but will also be particularly useful as a benchmark against which to justify the validity of the new system and evaluate its subsequent use.

Prototyping

A prototype, at this stage, must be seen as a further communication device intended to test the early concept with a view to identifying flaws and problems. As such, it only needs to provide a limited representation of the design to ascertain how suitable it is likely to be for the targeted users. These low-fidelity prototypes (Preece et al 2002) are best kept simple and easily produced, often resorting to paper-based versions of the conceptual design, to explore quickly and cheaply with other designers and end-users ways of improving the proposed design. Sketches, in particular, can provide the ideal 'conduit from idea-in-the-head to idea-in-the-world' (Saddler 2001:17) to convey abstract concepts as well as user interface design features.

From theory to practice: the shaping of the physical design

Considering the physical design of an application means looking at all the concrete aspects of the interface in relation to the agreed conceptual model. However, there is no convenient defining line between both stages as the process is 'transformational' (Shneiderman 1998). Whilst it gathers its own design momentum, it might be necessary to go back to the conceptual design in order to reconsider some design decisions due to unforeseen constraints which have subsequently come to light. Indeed, it must not be forgotten that, since the process is inherently iterative, design issues will be addressed and revisited during the conceptual design stage as well as during the physical stage. Design decisions should reflect the need to adhere to previously identified requirements, meet established usability goals and abide by design principles and guidelines. In this respect, they must comprehensively cover all the considerations and features affecting the design, starting from the choice of platform or environment. Indeed, resorting to using the web as a means of delivery must stem from a design decision and not a personal fancy. For easier reference, the use of a formal and systematic approach is advised whereby each design decision is presented with its rationale stemming from the conceptual design, the requirement it addresses and the guidelines, which support it. See the model template from the LGU authoring design process below (Table 2).

Alternative designs or the use of trade-offs

Designing is about making choices and finding solutions towards achieving a close enough 'fit' (Levy 1997:163) between needs and capabilities. As such, the design process is often compared to a balancing act insofar as design principles, as well as different and sometimes conflicting requirements, be they instigated by users, functional constraints or educational considerations, might require the need for trade-offs. For example, integrating different types of interaction, such as navigational and directional, within the same interface, might lead the designer to reconsider the principle governing consistency at the expense of learnability. Similarly, functionality might have to be traded off against accessibility. For example, a greater multimedia input will have an effect on the memory requirement, which in turn will lead to a reduced download speed and limited network access. The conceptual model may also be purposefully or artificially limited by an early design decision or a technological impossibility. This happened in both cases in the LGU design project. In the first instance, the decision was made to ignore an author requirement to introduce assessment and scoring as this would have counteracted a student requirement specifying the exact opposite as well as introduced a contradictory tutoring element into the distinct tool role played by the interactive interface. However, since the requirement in question remained on the design agenda, a separate testing tool was designed and developed. In the second instance, both authors and students required the integration and interactive use of sound online but this request was rejected at the time, in 1998, as the available technology could not deliver this medium adequately. This has since been remedied with a revised version of the authoring tool, which underwent the whole design process to fully integrate and exploit direct online listening, recording and speaking. Refer to Hémard & Cushion (2002) for further details.

From design decisions to solutions: using a design decision template

The following table shows a typical presentation of a design decision and its corresponding design solution. It shows and strengthens the correlation between requirements, decisions, supporting design guidelines and ultimately design solutions, which as a result, can be rationally explained and formally justified without having to resort to personal reasons and idiosyncratic choices.

Table 2. Example of a design decision template from the LGU authoring tool design process

Project:	LGU Authoring tool Version 2
Design Decision:	Enable sound recording within the same text authoring environment.
Rationale:	Facilitate the authoring of web-based sound to improve the authoring tool's output and interactive scope.
Requirements:	Fully integrate text and sound authoring, digital sound authoring must be easier to use than analogue sound recording, sound authoring must be synchronised with text authoring.
Guidelines:	Learning strategy, task support, organisation, navigation, effectiveness, presentation, consistency, sound (see full guidelines in Hémard 1997)
Design Trade-offs:	To keep the sound recording functionality easy to use and within the same text authoring environment, remove the editing facility for existing sound recordings from the interface.
Design Solution:	Interface to display 3 frames: text with answers, sentence to be recorded and standard sound recording facilities (Play, Record, Save with back and forth functions)

Design supports

Usability principles and guidelines

High-level design principles are a useful design support helping the designer to focus on global aspects of the conceptual design. They are sometimes referred to as heuristics to indicate that they are merely pointers providing reminders of what to do and what to avoid when considered. These include recommendations such as 'know the users and their needs', 'ascertain the user-task feasibility and match', 'select an appropriate syntax and semantics', 'be consistent and clear'. Others are concerned with the importance of visibility, feedback and constraints to reduce errors. These have been extensively covered in the HCI literature (see Norman 1988, Preece et al 1994, Shneiderman 1998) but also, more specifically, in the CALL literature (see Allum 2001, Davies et al 1994, Hémard 1997, Laurillard 1993).

Whereas the role of high-level design principles is primarily to inform the conceptual design, usability principles and guidelines are more prescriptive, therefore, more explicit and can be used as referential criteria for evaluation purposes.

Guidelines for designing for the web

Specific guidelines for web design can be found in the HCI literature. Preece's taxonomy (2000:275), which groups guidelines for usability into three categories: *navigation*, *access* and *information*, is useful and easy to use. Adopting a slightly different approach, Nielsen (2000) emphasises content design, focusing on the legibility and accessibility of the textual and multimedia displays, as well as site design, paying particular attention to navigational capabilities. Likewise, Shneiderman (1998) presents design considerations on web-page design, which can also be used as guidelines. These concentrate on page display, hypertext structures, sequencing, clustering and emphasis for embedded objects, as well as access and graphical design. The latter, however, comes with a warning note stressing that if 'simple web sites can be created successfully with authoring tools, doing innovative and effective page layout for a large web site requires as much care and skill as does laying out a newspaper, magazine or book.' (577). Additional guidelines concerning web page design, structure and information retrieval can equally be found in Larson & Czerwinski (1998).

Evaluation

> *If theory is not present explicitly in design, then we must ask on what basis, and with what criteria, the design will be evaluated. (Levy 1999:94)*

The importance of a theoretical basis to underpin the whole of the design process cannot be overstressed. By promoting a rigorous and principled approach to design, such a basis will facilitate the process of understanding users' needs and satisfaction, and help to generate a successful design solution to an identified problem. The role of evaluation is to ensure that this understanding is constant. Thus evaluation is cyclical. The greater one's appreciation of the needs of users, the more likely one's design will mirror this in the solution it provides. By the same token, the more acceptable one's design is to the users, the more relevant and pertinent their feedback will be. This, in turn, will facilitate the refinement and improvement of the design as part of the iterative process.

Again a number of methods exist (see Karat 1997), producing a range of quantitative and qualitative feedback data, which will be more or less applicable depending on the context of use, the types of users and the data required. The best evaluative methods that we have used in the context of online CALL have capitalised on the dual language teaching and designing role played by the CALL author. User proximity tended to favour the use of informal feedback and real life observation involving users, in this case teachers and students, observing what they did and getting immediate feedback from them. Another convenient and practical method, called predictive evaluation method, has also been systematically used. It is based on the use of a checklist of tried and tested heuristics closely related to design guidelines (see Table 3). Finally, a method worth considering is the user walkthrough (for further details see Hémard 1999, Hémard & Cushion 2001) for it generates the most comprehensive and accurate qualitative as well as quantitative data. In any case, it is advisable to resort to a range of different methods, since, no matter how good the testing of the product is, bugs, inefficiencies and user preferences will only come to light in its mass use and be cross-checked through its most comprehensive and critical exposure. The table below shows some of these advantages and related drawbacks of available methods.

Table 3. List of evaluation methods

Methods	When to use	How to use	Type of data	Advantages/Disadvantages
Informal feedback	At any stage in the design process: formative or summative.	Seeks reactions from individual students/staff following interaction with prototype/product.	Qualitative	Quick, easy and flexible. Can produce some very valid feedback from some users which can be directly fed back into the design but data can also be parochial and misleading.
Real life observation	At a later stage in the process for summative evaluation.	Can easily be used in language class, with students trying beta version of new system.	Qualitative	Can be very interesting in terms of eye opener. Can also create different interaction bridging CALL activities and language teaching and learning.
Checklist	At a later stage in the process: essentially summative.	Seeks input from peers or experts in the field.	Qualitative	Does not involve targeted users, but not necessarily easier to arrange. Convenient but can lead to a self-congratulatory box ticking exercise.

Methods	When to use	How to use	Type of data	Advantages/Disadvantages
User Walk-through	At requirement stage and later for summative evaluation	Best used in a computer lab with audio or video recording facilities. Most profitable when students are paired thus triggering a triangular interaction with computer.	Qualitative	See Table1 related to mental models and needs
Usability testing	At later stage of design process to measure users' performance on specific tasks	In a formal context in a computer laboratory controlled by the evaluator	Quantitative	Valuable feedback when successfully carried out, but students can be unruly participants who are not necessarily willing to abide by rigid rules.

The checklist: a first step towards sound evaluation

Design guidelines as heuristics

Design guidelines presented as heuristics can be particularly useful for evaluating a new application. As such, these can form the basis of a valuable checklist complementing the summative evaluation of the newly designed system. Garzotto et al (1995) recommend that designers consider the following heuristics for hypermedia evaluation in order to improve usability: How *rich* and reachable is the information? How *easy* is it to use it in terms of accessibility and learnability? How *consistent* is the conceptual representation of similar and different elements? How *self-evident* is the meaning and purpose of what is represented? How *predictable* is the interactive outcome? How valid and *readable* is the application? How *reusable* are objects and operations? Consider also this set of ten usability heuristics proposed by Nielsen (2001), which raise questions as to whether the system: always keeps users informed about what is going on; speaks the users' language; supports user control and freedom with clearly marked 'emergency exits'; is sufficiently well designed to prevent errors from occurring; provides objects, actions and options which are always visible; is flexible and efficient, providing shortcuts for experienced users; adopts an aesthetic and minimalist design to increase visibility; helps users recognise, diagnose and recover from errors; and finally, provides help and documentation.

More specifically, the following heuristics stemming from the set of guidelines for web design presented by Preece et al (2002:415) are particularly practical and efficient when applied to existing websites. Additional comments have been made to improve their applicability.

Ensure good navigation by checking the following:

- Are there any orphan pages? These can only lead users into frustrating dead ends.

- Are there any long pages? Scrolling slows navigation and can be tiresome and irritating.
- Is the necessary navigation support provided? Users can get lost if a site map cannot be accessed at any time.
- What is the menu structure? Broad shallow menus are more usable than a few deep hierarchical ones, which can lead to serious disorientation.
- What is the colour used for links? Blue is the standard colour. If a non-standard colour is used, is the hyperlink obvious to users?
- Are menus and links consistent? Navigation and information will be enhanced by a consistent design approach.
- Is the architecture meaningful? Hyperlinks, by dint of being easily created, can proliferate and lead to disorientation.
- Is the architecture tailor made to the users' needs? Introducing patterns and link hierarchies can increase interaction by promoting different interactive paths.

Ensure accessibility by checking the following:

- How simple are the URLs? The more complex, the easier it will be to make mistakes.
- Are there many graphics, which can slow the download time? Slow downloading is frustrating and can even be a deterrent for users.
- How compatible is the site with existing browsers? Will it work on all the available versions?

Ensure that the design of the information is appropriate by checking the following:

- How is the page layout? Is it well structured? Is there too much text? By being well presented, a website will make a greater impact on users and will be easier to understand. It must also be kept in mind that 'breaking a text into linked fragments does not ensure that the result will be effective or attractive' (Shneiderman 1998:556). For further information, check the Golden Rules of Hypertext (Shneiderman 1989).
- Is the information complete, up-to-date and, we would add, accurate? An incomplete, outdated and inaccurate website will lose credibility with its users.
- How are colours used? Are they used excessively or inexplicably? Are they garish? Are they the result of a personal choice? Are they used to define specific roles and functions or are they merely used to make the site attractive? Even if colours stem from a personal choice they must always be justified.
- Are there graphics and animations? Are they indispensable to the design or interaction? What are their added values to the site? Personal satisfaction in authoring flashing banners and animations may not be shared by users and may even become a source of irritation when their use or display is systematic (see Felix this volume).
- Are all the buttons, fonts and menus used in a consistent manner throughout the site?

User acceptability

Ultimately, a successful interface will not only meet identified user requirements and have achieved its usability goals to be usable but it will also need to become acceptable to the users. Indeed, users will need to recognise and relate to the general principles governing the design of the interface. Furthermore, they must perceive benefit from interacting with such an interface and feel empowered by it in order to find satisfaction using it. This elusive concept of acceptability, key to a successful interaction and integration, will only come into play if the design process is user centred and thus prioritises usability over functionality. This is all the more relevant for online CALL, whose successful accessibility must rest on users' willingness to interact with it and use it as a valuable learning support. On this premise, online CALL activities must not only be useful and meet students' needs but must also be sufficiently enjoyable to be accessed outside the classroom. Thus, it is only when the overall validity of online CALL applications becomes openly or tacitly acceptable to the users that further progress will be made and better feedback obtained. In particular, this necessary student satisfaction will be a key factor when considering new learning outputs and performance, which, in turn, will provide valuable evidence that online CALL can make an impact on the user population, affecting both attitudes and working practices.

Conclusion

If the design process is inherently iterative and evaluation cyclical, so too is CALL development. The greater the feedback, the more we will know and understand about users and the more acceptable the CALL interface will be to them, generating further design improvement and wider access. By presenting a design process supporting a best practice model, with references and illustrations, it is to be hoped that this will promote an informed view of online CALL design and contribute to such a necessary block-building design development. Finally, by dovetailing theory and design, combining a top-down conceptual approach and a bottom-up evaluative process, online CALL authors will be able to better reflect on their design decisions whilst being in a much stronger position to justify their design solutions. In turn, this will help them withstand closer scrutiny and generate appropriate design modifications towards greater usability and acceptability.

References

ACM SIGCHI (1992). *Curriculum for Human-computer Interaction*. ACM Special Interest Group on Computer-Human Interaction Curriculum Development Group, New York.

Allen, R. B. (1997). Mental Models and User Models. In M. G. Helander, T. K. Landauer & P. V. Prabhu (Eds.), *Handbook of Human-Computer Interaction*, 49-65. North-Holland: Elsevier.

Allum, P. (2001). Principles applicable to the production of CALL-ware: learning from the field of Human Computer Interaction. *ReCALL*, 13 (2), 146-166.

Bainbridge, L. (1990). Verbal Protocol Analysis. In J. R. Wilson and E. N. Corlett (Eds.), *Evaluation of Human Work: A Practical Ergonomics Methodology*, 161-179. London: Taylor and Francis.

Bickerton, D., Stenton, T. & Temmerman, M. (2001). Criteria for the evaluation of Authoring Tools in language education. In A. Chambers & G. Davies (Eds.), *ICT and Language Learning: A European Perspective*, 53-66. Lisse: Swets & Zeitlinger.

Cameron, K. (Ed.) (1999). *Media Design and Applications*. Lisse: Swets & Zeitlinger.

Davies, G., Hickman, P. & Hewer, S. (1994). *Style Guidelines for Developers*. Hull: The TELL Consortium, University of Hull.

Davies, G. (2001). New technologies and language learning: A suitable subject for research? In A. Chambers & G. Davies (Eds.) *ICT and Language Learning: A European Perspective,* 13-29. Swets & Zeitlinger.

Erickson, T. D. (1990). Working with interface metaphors. In B. Laurel (Ed.), *The Art of Human Computer Interface Design*. Boston: Addison-Wesley.

Felix, U. (2001). *Beyond Babel: language learning online*. Melbourne: Language Australia Ltd.

Garzotto, F., Mainetti, L. & Paolini, P. (1995). Hypermedia design, analysis and evaluation issues. *Communications of the ACM*, 38 (8), 74-86.

Hémard, D. (1997). Design Principles and Guidelines for Authoring Hypermedia Language Learning Applications. *System,* 25 (1), 9-27.

Hémard, D. (1998a). *Theoretical Framework for Authoring Hypermedia for Language Learning*. Unpublished doctoral dissertation, London Guildhall University.

Hémard, D. (1998b). Knowledge Representations in Hypermedia CALL Authoring: Conception and Evaluation. *Computer Assisted Language Learning*, 11 (3), 247-264.

Hémard, D. (1999). A Methodology for Designing Student-Centred Hypermedia CALL. In R. Debski & M. Levy (Eds.), *WorldCALL: Global Perspectives on Computer-Assisted Language Learning*, 215-228. Lisse: Swets & Zeitlinger.

Hémard, D. & Cushion, S. (2000a). Authoring a Web-enhanced interface for a new language-learning environment. *ALT-J, Association for Learning Technology Journal*, 8 (1), 41-49.

Hémard, D. & Cushion, S. (2000b). From Access to Acceptability: Exploiting the Web to Design a New CALL Environment. *Computer Assisted Language Learning*, 13 (2), 1-16.

Hémard, D. & Cushion, S. (2001). Evaluation of a Web-based Language Learning Environment: the importance of a user-centred design approach for CALL. *ReCALL*, 13 (1), 129-142.

Hémard, D. & Cushion, S. (2002). Sound Authoring on the Web: meeting the users' needs. *Computer Assisted Language Learning*, (in press).

Johnson, J. & Henderson, A. (2002). Conceptual Models: Begin by Designing What to Design. *ACM Interactions*, January / February.

Kahn, P. (2001). Information Architecture: a New Discipline for Organizing Hypertext. *Proceedings of the 12th ACM Conference on Hypertext and Hypermedia*, University of Aarhus, Århus, Denmark, August 14-18.

Karat, J. (1997). User-Centered Software Evaluation Methodologies. In M. G. Helander, T. K. Landauer & P. V. Prabhu (Eds.), *Handbook of Human-Computer Interaction*, 689-704. North-Holland: Elsevier.

Larson, K. & Czerwinski, M. (1998). Web page design: implications of memory, structure and scent for information retrieval. Proceedings of *CHI 1998*, 25-32.

Laurillard, D. (1993). *Program Design Principles.* Hull: TELL Consortium, CTI Centre for Modern Languages, University of Hull

Levy, M. (1997). *Computer-Assisted Language Learning – Context and Conceptualization.* Oxford: Clarendon Press.

Levy, M. (1999). Design Processes in CALL: Integrating Theory, Research and Evaluation. In K. Cameron (Ed.), *Media Design and Applications* (1999), 83-107. Lisse: Swets & Zeitlinger.

Lewis, C. & Wharton, C. (1997). Cognitive Walkthroughs. In M. G. Helander, T. K. Landauer & P. V. Prabhu (Eds.), *Handbook of Human-Computer Interaction*, 717-733. North-Holland: Elsevier.

McAleese, R. (1999). Navigation and Browsing in Hypertext. In R. McAleese (Ed.), *Hypertext: Theory into Practice*, 5-38. Exeter, UK: Intellect.

Nielsen, J. (2000). *Designing Web Usability.* Indianapolis, USA: New Riders Publishing.

Nielsen, J. (2001). *Ten Usability Heuristics.* http://www.useit.com/papers/heuristic

Nielsen, J., Coyne, K. P. & Tahir, M. (2001). Make it Usable. *PC Magazine.* http://www.useit.com/papers

Norman, D. (1988). *The Design of Everyday Things.* New York: Basic Books.

Preece, J. (2000). *Online Communities: Designing Usability, Supporting Sociability.* Chichester, UK: John Wiley & Sons.

Preece, J., Rogers, Y. & Sharp, H. (2002). *Interaction Design: beyond human-computer interaction.* New York: John Wiley & Sons, Inc.

Preece, J., Rogers, Y., Sharp, H., Benyon, D., Holland, S. & Carey, T. (1994). *Human-Computer Interaction.* Wokingham, England: Addison-Wesley Publishing Company.

Saddler, H. J. (2001). Understanding Design Representations. *ACM Interactions*, July-August 2001.

Shneiderman, B. (1989). Reflections on authoring, editing and managing hypertext. In E. Barrett (Ed.), *The Society of Text*, 115-131. Cambridge: MIT Press.

Shneiderman, B. (1998). *Designing the User Interface; Strategies for Effective Human-Computer Interaction.* Reading, Massachusetts: Addison-Wesley Publishing Company.

Veen, J. (2000). *The Art and Science of Web Design.* Indianapolis, Indiana: New Riders.

Vora, P. R. & Helander, M. G. (1997). Hypertext and its Implications for the Internet. In M. G. Helander, T. K. Landauer & P. V. Prabhu (Eds.), *Handbook of Human-Computer Interaction*, 877-914. North-Holland: Elsevier.

All websites in this chapter were verified on 18.09.2002.

3

Optimising web course design for language learners

Robert Godwin-Jones, Virginia Commonwealth University, U.S.

Introduction

Much of what is being used today in computer-assisted language learning (CALL) is created with general purpose software and tools rather than with programs designed specifically for language learning. Language learners are likely to spend a good deal of any dedicated computer use time with language proofing tools of generic word processing programs or accessing web-based sources using a standard web browser. To be sure, dedicated language learning software continues to be developed and used. But issues of access, cost, and convenience lead many language teachers to use technology tools already installed and available to them and their students. Also, technology use in teaching and learning has moved beyond early adopters to the mainstream; while CALL enthusiasts might seek out new and innovative language learning software, that is less likely to happen with instructors for whom technology use is not a high priority.

Obviously, general-use software and tools are not developed specifically with language learning in mind. Consequently, their suitability for CALL varies. As Carol Chapelle points out, this necessitates a discriminating analysis of how such products can best be used: 'There is no shortage of general-purpose authoring and computer-mediated communication software from which some types of CALL activities can be constructed. What is needed are theoretically and empirically based criteria for choosing among the potential design options and methods for evaluating their effectiveness for promoting learners' communicative L2 ability' (Chapelle 2001:42). It is important to look at not just *what* is being used for CALL, but *how* it is being adapted for language learning. We will be examining web course site creation from this perspective.

The typical starting point for language teachers interested in using the web in teaching has been the creation of a personal home page and/or a links page for their students. As more resources are added to such pages, they have tended to evolve into course web sites, which incorporate a variety of materials and tools for the use of specific groups of students or subject areas. Initially what was included depended on the computer savvy of the teacher creating the site since the individual components needed to be created and linked by the instructor. This necessitates at a minimum a basic knowledge of web authoring and the HTML language and might involve – for creation of interactive exercises – mastery of a programming or scripting language such as Perl or JavaScript.

In the mid to late 1990s, a new option emerged for teachers – use of a template-based web course authoring and management system, often called a Learning Management System or LMS. Initially such tools were locally created and limited in scope, with simply the purpose of allowing content to be placed into web pages without the knowledge of HTML. As they have evolved into commercial products, considerably more functionality has been added (and continues to be added). Today products such as *Blackboard* and *WebCT* provide a rich array of web-based content creation/management tools and user interaction functions. They are being widely used internationally in secondary schools as well as in higher education. Institutions which invest in these products typically provide considerable training and support for faculty to use them in creating course sites, often to the exclusion of other options. For many language teachers wanting to create web sites, use of *Blackboard*, *WebCT* or another LMS provides the only realistic option. Consequently we will include consideration of their use in creating sites for language classes. Our fundamental purpose, however, will be to arrive at a set of best practices to recommend in developing course web sites, whether they be created with an LMS or not. Our discussion will center around three principal areas: *the course environment*, *content creation*, and *communication and interactivity*.

The course environment

Course web sites normally have a home page as a starting point for accessing the web resources available to students. Course sites might consist of only a home page, with an electronic version of the syllabus, an array of web links, or a list of readings and requirements. Some are designed simply as a means to provide information about the course for potential students. Others are intended to provide a rich array of services to enrolled students: day-to-day schedules; special announcements of event or schedule changes; access to handouts, worksheets, sample exams, or bibliographies; checking of individual grades or current grade average; links to web sites containing information which elucidates or extends content covered in the course. Course web sites may offer interactive functions such as electronic assignment submission, practice exercises, and online quizzes. Electronic communication is generally enabled through the availability of discussion forums, chat rooms, or e-mail links.

Which of these tools and services instructors elect to make available to students depends on a number of factors. If the web site is simply a companion to a classroom taught course, it is likely to be less extensive than a stand-alone web course. Distance education courses taught over the web will normally include a rich set of communication tools, extensive structured content, and a means of electronic assessment of student performance. The scope of companion web sites varies considerably depending on logistics such as available tools, technologies, and training; technical skills and interest of the instructor; and the nature of the course being taught. Some discrete point grammar and vocabulary drills might be important in a course site for a beginning language class, whereas discussion forums might form the nucleus of a site for an advanced culture or literature class. Inclusion of sound files may be of central importance in a phonetics course, while electronic chat might be especially beneficial to a conversation class.

Depending on the number and nature of elements of the course, site organization and navigation may be an important design consideration. A limited site might have a quite simple structure, with just a page linking all the course resources. To return to the course home page, students click the browser's back button or, preferably, return buttons included on the content pages. It is helpful for all pages to include identifying information such as date created/modified, e-mail link to instructor (or site maintainer), and an up-link to departmental or school home pages. More extensive sites benefit from a more highly structured approach, with separate pages dedicated to areas such as web links, assignments page, or communication tools.

One of the benefits of using a learning management system is that such a structure is already in place. A course shell is provided into which content and tools can be added. As is necessarily the case with any template-based system, options are fixed and customization is limited. The emphasis is placed on ease of use, to enable any instructor to create a course site, without prior knowledge of web authoring. The basic course layout used by *Blackboard* and *WebCT* is similar and typical of LMS products, a frames environment, with navigational links placed on the left side of the screen and a 'bread crumb' trail to show users their current location within the course. The course home page provides the starting point for navigating to all course resources. By default, *Blackboard* initially displays the announcements page, as shown in Figure 1, although individual instructors may choose a different initial view.

Figure 1. Announcement page within a Blackboard environment

WebCT displays a set of icons representing the main course areas as its initial view; they can be customized considerably by the instructor. *WebCT* offers the option of icon titles rendered in several West European languages, a feature lacking in the current version of *Blackboard*. However, neither system is a good choice if the intent is to provide an immersion-type experience in the target language (unless it is an ESL class). Both companies have promised to deliver more completely localised versions in the near future.

A frames environment is used by *Blackboard* and *WebCT* in order to provide a navigational bar which is always available to the user. This works effectively but does limit the display of content in the main frame, which may be an issue in some instances. The frames environment, as well as the use of Java and JavaScript (particularly in *Blackboard*) raise accessibility issues, although both products have improved recently in this area. The presentation of the material and the general look and feel of the web site can only be modified slightly – the basic structure of the site remains the same. While this uniformity cheers the hearts of IT managers and university administrators, whose lives are thereby simplified, it might not provide the best organization for specific needs or purposes of individual instructors. There may be pedagogical reasons for different site structures. Teachers might want to design a site around a theme such as taking a trip, creating a product showcase, or playing a game show. Other teachers might choose a multimedia approach, with audio-based navigation. A student-created structure is another possibility, with the organization of the site determined by student contributions.

An example of an alternative organisation of a course web site is described by Dorothy Chun and Jan Plass for an intermediate German class (Chun & Plass 2000:154-159). The scenario they have developed for the *netLearn* environment is that of students preparing to study abroad at a German university. Students make arrangements for applying for admission through the web site, as well as using it to arrange housing (by accessing actual housing ads from current German media sites), and to make travel arrangements. Their first task is to read (or listen to) a letter of invitation in German. For this and other tasks, glosses are available, as well as access to a German-English online dictionary. As they carry out their tasks on the web site, students create documents including text, audio or graphics, all of which are stored in the students *netPack*. The *netPack* and other tools are available at all times from a navigational bar.

The *netLearn* site uses a constructivist approach to how students learn through the web site (Chun & Plass 2000:154*ff.*). Students are assigned certain tasks, but they choose their own paths and the order in which they are done. Some of the material is also available in multiple formats, as text, audio or video. The *netPack* students create becomes a portfolio record of student performance. It would be difficult to create such a site in an LMS, which are only customizable up to a point. In *Blackboard*, for example, it is possible for individual instructors to choose to hide particular navigation buttons, but not possible to add their own. Thus ubiquitous access to instructor added tools is not possible. Student portfolios are also not a current feature of *Blackboard* or *WebCT*.

Another alternative scenario for a course web site is that created by the author for an intermediate German class. The principal readings in the class were excerpts from classic German children's books, Wilhelm Busch's *Max und Moritz* and Heinrich Hoffmann's *Struwwelpeter*. Navigational icons for the site were created from the graphics used in those stories, while the page background was designed to imitate parchment in color and texture. The goal, as shown in Figure 2, was to create a common look and feel to all course pages, using a unified theme which is carried over to the content as well (reading, discussion topics, online games).

Figure 2. Course homepage designed by the author

Course Home Page -- **German 201** (section 2) -- Fall, 1995		
General Info	Classwork	Interactions
Instructor & Access Info Policies & Procedures	Course Schedule Assignments	Students - Kaffeeklatsch Stammtisch - Ausbruch!
Study & Master	Read-View-Listen	Apply & Explore
Grammar Tutorials Exercises	On-line Stories Vocabulary Exercises	Internet Resources For German
Language Courses - Foreign Language Department - International Trail Guide - VCU		

A central activity of the site is a game entitled *Ausbruch* ('Breakout'). The students were given the following scenario: the mischievous children from the stories have mysteriously re-appeared as real people in present-day Germany and are wreaking havoc. Students are to find their whereabouts. They work in groups to solve clues by looking for information in short readings online (also by Wilhelm Busch) and from current German web sites. Once a group has discovered a clue, it is entered into a web form, which when submitted with the correct information, returns a piece of the puzzle needed to solve the case. Working with *Ausbruch* requires students to communicate with each other in the target language, and retrieve information from authentic language materials. The server-based scripts that run *Ausbruch* are not functions which could be created with tools available in an LMS. Of course, such a game could be created and then linked into an LMS, but it would be not as thematically or logically integrated into the course site.

Anyone creating a custom interface for a course site must ensure that sufficient navigational guidance is provided. Using an LMS provides a safe solution in that students will likely already be familiar with the site conventions and be able to navigate with little difficulty. Other built-in features of an LMS would be difficult to reproduce independently. An important feature is security. Students are assigned login ids and passwords and must use these to enter the course web site (although optional guest access can be enabled). Because a student is identified when the site is accessed, the system can also track what parts of the course the student visits. This monitoring results in the availability of detailed reports on student use of the site. This can be quite helpful in ascertaining use of resources generally and also by individual students.

Content creation

It has become an accepted tenet of second language instruction that learners need to work with linguistically and culturally valid texts (Omaggio-Hadley 1993). The web supplies such authentic language materials in abundance, despite the predominance of English on the Internet (Shetzer & Warschauer 2000:171). Teachers today are able to provide to students a variety of up-to-date materials as diverse as family histories, government studies, train schedules, company reports, bank deposit forms, travel journals, movie reviews, job ads, and newspaper articles. What makes such materials particularly enticing for language study is that they have become part and parcel of how cultures interact and communicate. Tapping into such resources can involve not just learning about the culture, but actually participating in it. Of course, the challenge is finding the resources at the appropriate language level, insuring the viability of the site, and turning the web document into 'comprehensible input' (Krashen, 1982:9). There should also be a specific task associated with accessing the web site, such as gathering a particular piece of information (for example, a recipe for *Apfelstrudel*, times of train connections between Munich and Berlin, current values of Volkswagen shares on the Frankfurt market), submitting it to the instructor, or even better using the information retrieved in a transformational way (baking the *Apfelstrudel*, making a train reservation, using a broker simulator to buy or sell Volkswagen shares).

The initial goal for many teachers in setting up a course site is often to provide a jumping off point for web-based assignments or a resource depository for projects. Given the vagaries of web searches and the proliferation of web pages of questionable quality, providing an initial orientation through a set of selected web sites can provide a valuable service to students. The links page can be even more useful if the web sites are briefly annotated, providing students with information to make web mining more fruitful and efficient. Teachers whose students access their pages from controlled environments such as classroom computers or language labs have also used web pages to incorporate links to external programs already loaded on the computer, such as courseware or multimedia learning programs. This provides an integrated resource collection of local and online resources which enables quick and easy access for students.

An option for the links page is to allow students to add recommended sites. We have used this function several times in course web sites and found it to be most useful in more advanced courses, in which students can also add comments or a description. This moves the links page from being a static to a collaborative page and involves students in building the course web site. In a recent German media course, students added links to web sites dedicated to German punk and hip hop bands, which other students found quite compelling, but which were not sites we would have chosen (or found easily).

Creating a page of web links is easily done in an LMS, or, for that matter, in a home-grown site. When working with an LMS, the option is provided to link an outside web site in a new external browser window or within the main content frame of the course site. To maintain consistency with the navigational structure of the site, most instructors elect to display linked sites within the frames environment. For many sites this provides an acceptable view of the site, however many web pages today are designed for full browser window display and may require too much scrolling in a frames environment. Consistency is recommended, whichever method is used, so that students know what to expect when going to access external web sites.

Teachers sometimes find web sites which provide just the information they are seeking and want to make available to their students. Then one day the site is suddenly gone, with no forwarding address. One way to forestall this kind of frustration is to load a web site locally. If asked, most maintainers of web sites do not object to such local caching, once they understand that is for limited educational not commercial use. Recent versions of Microsoft's *Internet Explorer* allow saving web pages as 'web archives' which save not just the text, but any graphics and external scripts used on the page as well. Using a saved web site is not as optimal as accessing a site live, but it can at times be valuable, especially if a lot of work has been done building a lesson plan around a site.

Most instructors are likely to link not only external web sites and resources but to upload and link instructional materials created themselves. In fact, one of the powerful features of the web is the ability to link seamlessly local materials with those on sites throughout the world. Instructor materials can be easily uploaded in an LMS and may be maintained and linked in its native file formats (such as Microsoft *Word* or *PowerPoint*) or converted to text or HTML. Linking files with proprietary formats necessitates student access to the applicable programs or viewers. Depending on the browser used, the display of such files is within the browser window or in a newly spawned external window. Many instructors simply maintain handouts or worksheets they want to make available in their original formats. However, most recent versions of word processing and other programs allow for easy conversion to HTML, which in some cases may be preferable. Conversion to HTML means that any browser can display the content without the need for a special viewer or the originating program (potentially a browser on a portable Internet device). More significantly, HTML versions of content allow for integration of learning tools which the LMS may make available. *WebCT*, for example, enables users to add simple hypertext glosses and quick comprehension checks (true-false or multiple choice questions) to uploaded content.

It is also possible to create comprehension aids for language materials such as short texts outside of the LMS and then link or import them into the course web site. SLA research points to the usefulness of glosses for learners (Chapelle, 2001:71). The page in Figure 3 shows a text used in a German media class. A word list was generated from a concordance, then keywords were extracted after a filter deleted common vocabulary. A glossary file was created and linked to the text file (through JavaScript). Students can show (or hide) glosses in a variety of ways, through font changes (bold, italics, colored), as hypertext links, or through pop-up windows. Students can click on any word or highlight a phrase. If the selection is in the text gloss it is displayed, otherwise an online German-English dictionary is searched.

Figure 3. Text comprehension aid

The principal on which this is based is that of user-requested help (Chapelle, 2001:72); the intent is to encourage students to work with the material on their own, using the glosses only as needed. This page is loaded into the *Blackboard* course site and accessed in that way by the students, but it could also be used independently of the course site. When linking instructor created pages instructors have the option of displaying the content within the frames environment or in a new window. Instructors intending to integrate a resource into the LMS can take the frames environment into consideration when designing the page.

Both major learning management systems allow content to be created online by entering (or copying) text into text entry fields. *Blackboard* also enables the creation of 'learning units', intended to provide a means for sequential content presentation. Neither *Blackboard* nor *WebCT*, however, provides the equivalent of an intelligent tutoring system, in which checkpoints can be set to allow or disallow a student from moving through content until mastery is verified (by a comprehension assessment) or providing content branching on student choices or student performance (see Arneil & Holmes, Bangs this volume). Nor do they provide the auto-resume function of such a system, in which students are returned to the point at which they left the program the previous session. Typically such learning modules are created with sophisticated authoring software such as *Director* or *Authorware* (both from Macromedia). It is possible to link a module of this kind into an LMS. The main course content for distance education courses marketed by *Class.com*, for example, is created in *Director*, then linked into *Blackboard* sites. However, the interface and navigation for the content modules are quite different from those of the *Blackboard* site. *Blackboard* has begun a 'Building Blocks' initiative which allows third-party tools to be integrated as plugs-in into the *Blackboard* environment. Several content creation and management tools are available as plug-ins, which have the advantages of greater ease of use and compatibility with the *Blackboard* user interface.

The challenge of rendering online material accessible to students may be even greater for multimedia resources than it is for text. There has been in recent years an enormous growth in the volume of digital audio and video materials on the web in a variety of languages. Internet radio stations abound and many broadcasters make clips available for instant listening and viewing. Web-based multimedia can significantly enrich a language learning environment, but its use also poses special problems, technical, linguistic, and pedagogical. Users must have the appropriate players or plug-ins for the media being run. The clips are often highly time-sensitive and may not be available for long periods of time. Depending on the level at which such resources are being used, comprehension aids may also need to be developed. Transcripts are normally not available; transcribing even short audio or video passages is a time-consuming and arduous task.

Figure 4. Audio comprehension aids

Vorstellungsgespräch Verständnishilfe	Stichwörter
Herr Geibel Nein, nein, ich kann gar ... Aber wenn da etwas fehlt, ich sage ja nur: wenn ... Sie können dann gern ... außerhalb der Arbeitszeit einen ... besuchen. Wir tragen die Kosten, Sie bringen die Zeit ein. So ist das bei uns die Regel. Also, ... das. Aber noch ... in diesem Zusammenhang: Haben Sie auch ..., kennen Sie auch das Land und die Leute, die ...? Im ... habe ich von Frankreichaufenthalten gar nichts gefunden. Deshalb ... ich.	**wenn da etwas fehlt,** - if there is a problem there **auf Firmenkosten** - paid for by the company **überlegen** - consider **Landeserfahrung** - experience in the country **Frankreichaufenthalt** - visit to France **Deshalb** - therefore

Play/pause => ( )   **Help Options ->** ☑ Show Keywords   **Transcript:** ○ None   ◉ Partial   ○ Full

No plug-in? Play audio through external player or as MP3 file. Use these buttons to move manually through the help screens: [<=] [=>]

Figure 4 shows how playback of an audio file has been supplemented by comprehension aids. In this instance, students are encouraged to listen to the text first without any aids, then to select from the available aids of keyword listing, partial transcript, or full transcript, depending on the desired level of help. The help displayed is synchronised to the audio and updated dynamically. Students may change help options on the fly as the audio plays or can move through the help screens manually. In this instance, the audio is delivered in RealMedia format; it is advisable to provide several media formats for media, such as here the additional MP3 versions. Streaming media such as RealMedia, MP3, or QuickTime are preferable over audio formats such as Wave and AIFF which must first be downloaded in their entirety before playing. This page can be used equally well in a self-created web environment or within an LMS. LMS allow for uploading and linking of audio or video files (or access from a CD or DVD), but do not provide a means for creating comprehension aids for media playback. When using audio or video, it is recommended that a folder be set up on the course web site with links to media players and other utilities available on the site.

Communication and interactivity

In addition to serving as a rich repository of language documents and realia, the web also offers communication and collaboration opportunities which enable a variety of options for language use and practice. Language teachers discovered early on the benefits of both synchronous communication such as chat rooms and asynchronous communication such as discussion forums (Warschauer 1995). Both offer the potential for learner-learner as well as learner-native speaker interactions. Studies have found that the use of discussion forums typically results in wider participation by all learners and more opportunities for writing than is the case in a typical classroom environment (Warschauer 1997).

Both *Blackboard* and *WebCT* provide tools for student-student and student-computer interactions. One of the most widely used are the discussion forums. Once again, for some instructors, the ease of creating and using multiple discussion forums for a class provides the main rationale for creating a course web site with an LMS. LMS forums have a number of useful features including varied display of message lists (for example, hiding already read messages), inclusion of attached files, and restriction of forum access to selected class members. One of the consequences of the controlled access to LMS sites is that forum participation across classes or schools is problematic. Access can be granted to outside users, but this can be cumbersome. It may be more practical in such cases to exchange messages through e-mail, although the easily navigable overview and archiving of messages in discussion forums is sacrificed.

Many language teachers find the asynchronous nature of e-mail and online forums more practical than the synchronous chat tool in which all users need to be online at the same time. Of course, the nature of the writing tends to be different in the different media, with chats more closely resembling spoken exchanges (short, spontaneous, unstructured), while forum messages are more typical of written language (longer, more considered). There is evidence that in discussion forums students are more likely to take the time to reflect on form as well as meaning (Kern and Warschauer 2000:2). One advantage of chat is the possibility of including graphics in user messages, which may be useful in some instances. There are many stand-alone chat and discussion forum programs, which are available to instructors not using an LMS.

In addition to sharing information through forums or chat, it is also possible in an LMS for students to exchange files. While files can be attached to forum messages, another option is to use the file exchange system incorporated into the LMS. This is particularly of interest in *Blackboard*, in which groups can be set up in which certain students can exchange files as well as participate in group discussions. This feature is especially useful for group project work. The file exchange function also allows students to submit files for the instructor. It is possible as well for the instructor to return a marked or commented paper to the student. This provides a means to enable process writing in which student papers are improved through peer or instructor feedback.

Language teachers have also come to appreciate and use the capability of the web to create learner-computer interactions, mainly for the purposes of vocabulary learning, comprehension assessment, and discrete point grammar drills. Initially, such series were all server-based, with limited forms of feedback. Since the advent of client-side interactivity (mostly through JavaScript), more flexibility has been possible. Studies have shown that rich feedback to learners in such environments is beneficial (Chapelle 2001:21), and current web scripting options allow for quite sophisticated options, as well as fine-tuned parsing. Even more interactivity is available through the use of 'dynamic' HTML (DHTML). The power and flexibility of web interactivity can be seen in current versions of programs such as *Hot Potatoes,*

PracTest, or *Interactive Exercise Makers* (see Roever 2001 and Arneil & Holmes this volume).

Online assessments can be created with *Blackboard* or *WebCT*. These are self-correcting for formats in which there is a finite set of correct responses (true-false, multiple choice, ordering, matching, short answer). Essay-type questions can be assigned but must be individually read and marked by the instructor. The assessment systems in current LMS uses a server-based approach, so that feedback is supplied only when the entire quiz or exercise has been completed. Typically feedback is provided for right or wrong responses but not for anticipated incorrect answers. *WebCT* includes more feedback options, as well as more question types. It is possible to set up quizzes so that they can be taken multiple times or just once. Timed quizzes are also possible. Performance on quizzes is automatically added to a student's record in an electronic gradebook. However, because of security and identity issues, using online quizzes for formal assessment is not recommended.

One of the advantages of a server-based assessment set-up is the possibility of creating question pools, from which individual quizzes or tests can be generated with randomised selection of questions and a different mix of questions for each student. Another feature of this approach, as implemented in LMS, is the timed release of quizzes (and other content). Quizzes can be set up to be made available only during a specific period. The default for creating questions is text, but it is also possible to insert graphics, multimedia or a URL into a question, providing more flexibility in what kind of information forms the basis of the assessment. However, it is not generally possible to display content and assessment at the same time, for example, using one frame to display a short text or media file and the other to contain a set of comprehensive questions. In fact, the assessments are not easily used as learning tools, being instead designed for formal assessment. It is possible, as with other functions and tools, to import into an LMS quizzes created with a third-party tool such as *Hot Potatoes*. However, this poses the familiar problem of integration into the LMS environment. In this case also the gradebook is not tied to the online quizzes. While this may not be an issue for quizzes designed for self-assessment and as learning exercises, it may still be desirable for a record to be kept of which students took the quiz. It is likely that more interactive quiz creation options will become available as LMS plug-ins.

Issues in the use of learning management systems

Course web sites can provide a valuable service to students, by centralising a variety of electronic resources and tools which aid learning and communication. Instructors should keep in mind that if they expect students to use the web site on a regular basis, they themselves need to do so as well, keeping the content current and informative. Many instructors are finding that the ease of use of a learning management system expedites the upkeep of a web site and have gotten into the habit of updating sites on a regular basis after class (posting assignments, changing schedules, uploading handouts, and so on). They may find that a course site offers a convenient storage and retrieval system for their own content files. One of the advantages of using an LMS is the ease with which web sites can be recycled for use in subsequent semesters, with created, linked or uploaded content carried over to the new site. Uploading files without FTP is a key feature of LMS, although file management is course specific and does not duplicate the functionality of a digital library in terms of rights management, cataloging, or sharing. Also, some publishers are making textbook content available in formats easily importable into *Blackboard* or *WebCT*, including in some cases large questions pools.

Even technology-savvy instructors may find it useful to take advantage of the built-in tools available in a learning management system. While an LMS makes a large selection of tools available, instructors should not feel it is necessary to include them all in creating a course web site. Rather, the construction of the site should be determined by the nature of the course and the purpose the site will play. For example, setting up a discussion forum without specific directions for its use, may not be profitable. Instructors using an LMS are advised to search out services or add-ons that may be advantageous to language teachers. There is, for example, a plug-in from a third party (Wimba) which enables voice boards to be added to courses (see Felix this volume). *WebCT* has organized groups of instructor users by discipline. Advice and experiences from fellow language teachers who have used an LMS for their courses can be invaluable.

The number of language teachers creating course web sites with an LMS is likely to increase while fewer faculty will be creating web sites and resources independently. Language teachers should be aware, however, that there are drawbacks to using LMS for language learning. The method of content creation and linking does not encourage a multimodal approach, i.e., incorporating different versions (text, audio) of the same passage. We know from CALL research how important this is for some learners (Pusack & Otto 1997). SLA research tells us that specific language structures are developed in learners as they are needed (Chapelle 2001:46*ff.*), and that therefore it is best to provide students with a range of structures accessible in a flexible order. This kind of presentation also is problematic in the rigid hierarchy typical of the LMS environment. Work-arounds can always be found, but the interface does not easily lend itself to an alternative means of arranging materials.

The paradigm in an LMS is the student as recipient rather than as a seeker of knowledge and skills. The emphasis is on linking and uploading information, not on transforming information into learnable knowledge. Language learning is a complex process, not easily amenable to the limited structural options of an LMS. The simplicity of LMS is by design, with the intent of providing as shallow a learning curve as possible for users. The result is functionality that is quite useful but often fails to tap into the full power of the medium. One need only compare the 'learning units' in *Blackboard* to what hypermedia design research has shown about optimal courseware design (Grabinger & Dunlap 1996) to see how limited that functionality is in terms of what is possible and effective for the learning process (see also Bangs this volume).

Learning management systems tend to gather functions in separate groups, such as assessment or content creation or student tools. Cutting across these categories is not easy and in some cases not possible. This is especially the case in *Blackboard*, although this functionality is improved in version 6. Somewhat more inter-linking of resources is possible in *WebCT*. The structure of an LMS tends to lead instructors to think of web resources for their courses in pre-defined categories. While, as we have seen, it is possible to import content and functionality created outside of the LMS, experience indicates that once instructors begin using an LMS, they largely keep to the functions available within that system. In fact, that is one of the issues with the wide-spread use of LMS. The generation of teachers using these systems tend to assume that the functions and tools available to them are those possible to provide to their students. They necessarily buy into the mindset of the developers of the software. But in reality we have just begun to tap the enormous potential of web-enhanced instruction and it would be unfortunate if the popularity of LMS stifled the creativity and innovation language teachers generally bring to their task.

Conclusion

Warschauer & Kern (2000:12) maintain that networked computers are helping to create new paradigms of teaching and learning. But that is not likely to happen if teachers are not using the technology in new ways to explore undiscovered benefits. Teachers need to find ways to harness the knowledge they have gained through classroom practice to the use of network technology. While many second language classrooms are vibrant and unexpected, language web sites tend to be pedestrian and predictable. That need not be the case, whether one is using an LMS or not. Teachers might consider using the basic framework available in *Blackboard* or *WebCT*, but perhaps begin to explore alternate means of using the web through web resource creation with authoring software such as *Dreamweaver* or *FrontPage*, which allow users a full range of options for creating web content. LMS users should view the functionality available in the system they are using as a starting point, not the final word in creating a language learning environment for their students.

References

Chapelle, C. (2001). *Computer Applications in Second Language Acquisition: Foundations for teaching, testing and research.* Cambridge: Cambridge University Press.

Chun, D. & Plass, J. (2000). Networked Multimedia Environments for SLA. In Warschauer, M. & Kern, R. (Eds.), *Network-based Language Teaching: Concepts and Practice,* 151-170. Cambridge: Cambridge University Press.

Grabinger, S. & Dunlap, J.C. (1996). Links. In P.A. Kommers, S. Grabinger, & J.C. Dunlap (Eds.) *Hypermedia Learning Enivronments: Instructional Design and Integration,* 89-114. Mahwah, NJ: Lawrence Erlbaum Associates.

Higgins, J. (1988). *Language, Learners and Computers.* London: Longman.

Krashen, D. (1982). *Principles and Practice in Second language Acquisition.* Oxford: Pergarmon Press.

Omaggio-Hadly, A. (1993). *Teaching language in context.* Boston: Heinle and Heinle.

Pusack, J. & Otto, S. (1997). Taking control of multimedia. In *Technology-Enhanced Language Learning,* 1-46. Lincolnwood, IL: National Textbook Company.

Roever, C. (2001). Web-based Language Testing. *Language Learning & Technology,* 5 (2), 84-94.

Shetzer, H. & Warschauer, M. (1995). An electronic literacy approach to network-based language teaching. In M. Warschauer & R. Kern (Eds.), *Network-based Language Teaching: Concepts and Practice,* 171-185. Cambridge: Cambridge University Press.

Warschauer, M. (Ed.) (1995). *Virtual Connections: Online activities and projects for networking language learners.* Honolulu: Second Language Teaching and Curriculum Center, University of Hawaii.

Warschauer, M. (1997). Comparing face-to-face and electronic discussion in the second language classroom. *CALICO,* 13 (2&3), 7-25.

Warschauer, M. & Kern, R. (2000). Theory and Practice of Network-based Language Teaching. In M. Warschauer & R. Kern (Eds.), *Network-based Language Teaching: Concepts and Practice,* 1-19. Cambridge: Cambridge University Press.

Websites

Blackboard – http://www.blackboard.com/
Class.com – http://www.class.com/
Hot Potatoes – http://web.uvic.ca/hrd/hotpot/
Interactive Exercise Makers – http://lang.swarthmore.edu/makers/indexold.htm
WebCT – http://www.webct.com/

All websites cited in this chapter were verified on 11.10.2002

TOOLS

4

Servers, clients, testing and teaching

Stewart Arneil and Martin Holmes
University of Victoria, Canada

Introduction

The phrase 'online exercise', as used to refer to an interactive activity delivered through the World Wide Web or through a browser interface, actually covers two completely different types of delivery system:

1. Client-side self-access learning exercises (e.g. *Hot Potatoes* exercises)
2. Server-hosted database-driven exercises (e.g. *WebCT* or *Blackboard* quizzes)

Each of these is suited to different pedagogical purposes. The first type of exercise is appropriate for situations where learning, rather than evaluation, is the primary goal. It also suits delivery to an open audience, as in the case of a public website with open access serving a huge potential population of students (such as Bernard Dyer's *French Revision* site, aimed at high school French students anywhere in the UK school system). Standalone exercises in the form of HTML documents may be distributed over the web, an intranet, or on CD-ROM, for example as an adjunct to a textbook. The second type of exercise is typically part of a larger structure hosted by a Learning Management System (LMS) such as *WebCT* or *Blackboard*. It is generally used for a target audience of registered students who are required to log on, and exercises are viewed more as tests, with scores being stored for retrieval by teachers.

The first part of this article will examine these two paradigms in detail, explaining the advantages and drawbacks of each, along with simple explanations of the underlying technology from which the limitations and strengths result. We will also examine possible hybrid exercises that combine components of both. In the second part, we will go on to look at the larger picture, by addressing the emerging standards for the storage and delivery of standard exercise types, 'learning objects' and student data. Whether using client-side or server-side exercises, or more likely a combination of both, a choice will have to be made among a wide range of commercial and non-commercial products and systems for authoring and delivering content. Choosing suitable tools is essential if one's work is to be effective immediately, and remain useful in the long-term. We will discuss the importance of these standards, and their relevance to the average instructor or instructional designer.

Type 1: Client-side self-access learning exercises

Hot Potatoes exercises are a good example of this format. A typical scenario (graphically illustrated in Figure 1) would be as follows:

1. The exercise is downloaded to the student's computer from an external source (this would usually be a web server, but it could also be a network hard drive or a CD-ROM). The 'exercise' in this case consists of all the questions, answers and other data, as well as the code required to make the browser display it correctly, and to allow the student to interact with it.
2. The student works on the exercise inside the browser. There is no need for any further communication with the web server or other source of the exercise.

Figure 1. A type 1 interaction.

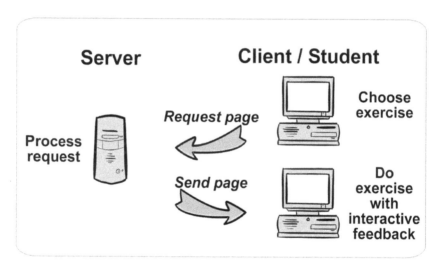

This scenario has several benefits from the student's point of view. Since all code is executed locally, the interface of the exercise (once it has downloaded) will be fast and responsive. It can also be highly sophisticated, analysing student input and comparing it with possible correct answers to provide helpful feedback during the learning process. This feedback can be immediate, in the sense that every time a choice is made or an answer entered, the exercise can respond instantly. This is less practical with feedback sent from a server, where the interaction would require communication to and from the server, as explained below. With instant feedback, features such as hints and clues are more effective, and the exercise can be more of a learning than a testing experience for the student; the object is no longer just to answer the question correctly, but to work towards a correct answer using a series of hints, clues and feedback. Students with intrinsic motivation, who enjoy discovery learning, may find this particularly appealing. (See also Bangs, this volume, for more on the distinction between exercises and tests.)

If the source is a CD-ROM, no network connection is even required, and huge data files (such as sound and video media) can be included if appropriate. Finally, in situations where network connection time costs money (such as in the case of a long-distance dial-up connection), the exercise can be downloaded, and then the client can be disconnected from the network for the length of time it takes to complete the exercise. A student might connect for two minutes to download an exercise, disconnect, and spend 30 minutes working on it without the need to communicate with the network. When the exercise is complete, and s/he wishes to move on to the next, then the connection could be re-established for another few minutes to access the subsequent materials. Finally, for the many situations in which interaction with the student is not wanted at all (such as when a publisher is distributing supporting exercises on a CD-ROM accompanying a textbook), the client-side exercise is the most reliable and cost-effective solution.

However, there are some disadvantages to this type of exercise delivery for tests as opposed to exercises. The most obvious is that, since all processing must occur on the client computer, all relevant information must also be delivered to the client along with the test. This of course includes all variants of correct answers. Although these might be encrypted, or be difficult to extract from the source code, nonetheless, the information is there for anyone who is determined to find it. What this means in practical terms is that there is no security, and therefore scores, even if they are recorded in some way, cannot be considered reliable, because there is always the possibility that the student achieved his/her score by hacking into the source code of the exercise. On the other hand, the issues of security related to any web-based delivery mechanism are huge; even if the exercise data remains secure on a server (as it might in the case of type 2 exercises, below), there is no guarantee of security anyway (see 'Security issues' below).

Although secure testing may not be possible, it is often useful to track students' scores and performance, both to help in assessing the student's progress, and to evaluate the quality of exercise material. This is not really possible in the case of 'pure' type 1 exercises (those in which there is no communication at all with the server after the initial download). This is because in order to save data, a file must be written somewhere by the computer; web browsers are typically not able (and should not be able) to write files on the client hard drive, since this would constitute a severe security risk for users browsing the web. If cookies are enabled, the browser can save small amounts of information, but this is not reliable, and in any case, retrieving the data even if it is saved can be difficult. In summary, we can say that 'pure' type 1 exercises cannot track student results or save them anywhere. This kind of feature might, however, be added to a type 1 exercise as detailed in the section below on hybrid exercise types.

Type 2: Server-hosted database-driven exercises

This type of exercise is most commonly encountered in the environment of Learning Management Systems such as *WebCT*, *Blackboard* and *LUVIT*. The typical scenario (graphically illustrated in Figure 2) is as follows:

1. The student logs in to the LMS system and to the course s/he is taking, and chooses an exercise.
2. The server creates an exercise from its database (possibly selecting questions at random from a large pool, or delivering a pre-built exercise). The exercise primarily consists of a form which must be completed (for example, by typing short answers into text boxes, or clicking on checkboxes or radio buttons to select multiple-choice answers).
3. The student completes the exercise. During this process, no feedback is provided.
4. When the exercise is complete (or possibly when a timer runs out), the exercise form is submitted back to the server, where the student's answers are compared with those in the database. At this point, depending on the level of sophistication of the server's quiz engine, some feedback may be harvested from the database based on the student's responses.
5. The server sends back a page containing a report on the student's answers. In the simplest case, this will just be a score; in the case of more complex systems, it may include detailed feedback on each of the student's answers.
6. The results are stored in a database for later retrieval by the instructor, and can be used to generate statistics for various purposes.

Figure 2. A type 2 interaction

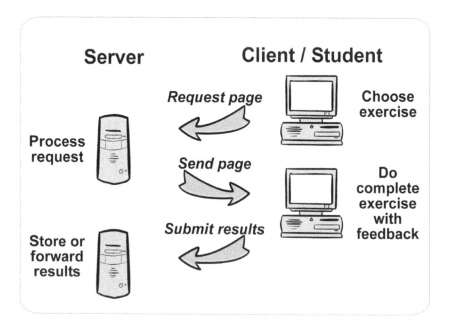

The differences by comparison with the type 1 exercise will be immediately apparent; first of all, the server does not have to send the answers and feedback out to the client, so some level of security is possible (although see the section on security below).

Scores, logon times, and all sorts of other data can be retrieved and stored for analysis, which can be extremely helpful, both for the instructor and for the materials designer. Finally, exercises can be created on-the-fly by selection from a large question pool. Doing this with a client-side quiz would entail transmitting the entire question pool to the client, which is inefficient, and may compromise the security of materials.

However, the process of doing the exercise is likely to be much less enjoyable for the student, because it is no longer interactive; the entire quiz must be completed before feedback is received. If the server does provide feedback, it provides a lot of it, all at once, and the student may already have forgotten the reason s/he chose a particular answer when doing the exercise, so the feedback may not be helpful. It is not possible to use the exercise as a learning experience in the sense of working towards a correct answer; the paradigm is closer to that of a formal test, where the score is more important than the process of doing the test. In essence, the quiz engines of most LMS systems are intended for diagnostic/testing purposes, rather than as instructional processes. Of course, such quizzes could be used as components of larger sequences of materials which have an instructional aim; for example, one could employ a traditional Test/Teach/Test model, in which the pre-test identifies instructional needs (as well as sensitising the student to the need for learning), and the final test confirms mastery.

It would be possible to code a server-side test so that hints, clues and feedback could be requested while the exercise was in progress; however, each of those requests would involve contacting the server, which would have to process the request and send the response back. The net effect of this from the student's point of view would be a sluggish interface; clicking on a Hint button might result in a wait of several seconds, which does not sound like a long time, but may be cumulatively discouraging, so students may stop asking for hints and help. In the absence of responsive interactivity, the exercises themselves tend to be pared down; instructors (or instructional designers) may be less inclined to spend time creating detailed hints, clues and supplementary material if they know that most students will probably be too impatient with the interface to use them.

It is important to note that the delay to which we are referring above is simply that due to network latency; this is not connected to the issue of immediate versus delayed feedback, in which *immediate* covers both the situations above, and *delayed* means that the program is designed such that all feedback is postponed until a designated later part of the process (such as when the student has completed a set of questions). For example, Pujolà (2001:88) discusses the difference between supplying feedback immediately on demand, and 'delayed two-step feedback', in which the student clicks once to receive information on the correctness of answers, and then again to receive explanatory material; this delay is built into the feedback system intentionally. In the German Tutor program used in Heift (2001), although not explicitly stated, we would guess that the feedback is produced on the server (since a parser is doing grammatical analysis of the students' input), but the question of network latency does not really arise, because students are submitting individual answers for analysis, and submission is the only way they can get any response (right/wrong + feedback). In other words, there are no optional clues or hints they could choose to use or ignore, so the sluggishness of the server response would not have any effect on their actions. Our point is that when there are feedback options such as asking for a clue (helpful information) or a hint (a part of the correct answer) which the student can use or not use, the tendency to use them would be much greater if the response is instant (client-side) rather than slightly delayed due to network latency (server-side). In addition, in the case of multiple-choice questions, we would argue that the tendency to gain extra reinforcement after correctly answering a question, by accessing the feedback for the wrong answers to find out why they were wrong, would be much stronger if that feedback is instant.

The final obvious disadvantage to this system is that a live connection must be maintained to the server throughout the process, and the materials must come from the server (in other words, they cannot be distributed through a static medium such as a CD-ROM, unless a quiz engine is also distributed on the CD or installed on the student's computer). Disconnecting from the server in the middle of the process may result in the submission being rejected, because the server may assume that something has gone wrong, or that cheating is taking place.

Type 3: Hybrid exercises

It should be possible to combine some of the features of both of the exercise delivery scenarios above to arrive at a system which is to some extent the best of both worlds. A number of options present themselves:

1. Client-side exercise pages (such as those produced by *Hot Potatoes*) can be uploaded as if they were ordinary content pages into an LMS system such as *WebCT*. This enables the LMS system to provide elementary security options such as password-protected login and access logs, while the exercises themselves, when accessed by the client, will function normally. Of course, this means that the score-tracking and record-keeping functions associated with the LMS system's own quiz engine will be unavailable; the server will be able to log the fact that a student accessed the page, but not what score s/he achieved.

2. Client-side exercises can be made to submit results to a server on completion of the exercise, by making a call to a server-side program such as a CGI script. *Hot Potatoes* exercises can be made to do this using the CGI submission option. In this scenario, the exercise is downloaded and done in the normal type 1 manner, but as soon as it is complete, the page sends the student name, score and timing data to a commonly-used PERL script running on a server. The script (*FormMail.pl*) then sends that information, in the form of an email, to a specified email address, presumably that of the instructor. Using modified code, *Hot Potatoes* users have also caused the script to save data in a simple flat-file database, so that statistics can be compiled.

3. A third option would be to create an LMS system which is designed specifically to serve client-side exercises. The authors are currently working with a UK company to develop such a system, called *hotpotatoes.net*. In this scenario, an LMS system is operating on the server (allowing student logins, tracking, class groups, etc.). Exercises are uploaded into this system directly from the authoring tools (in this

case *Hot Potatoes* or *TexToys*). On upload, the instructor can specify access controls. Students logging on to the system can access the exercises intended for them, and these exercises are downloaded to the web browser, and function in the normal way. At the end of the exercise, though, the student and performance data are sent back to the server, where they are stored in a database. The instructor can log on to the server and view or download that information, as well as managing resources and access as in a normal LMS system.

Figure 3. A type 3 hybrid exercise

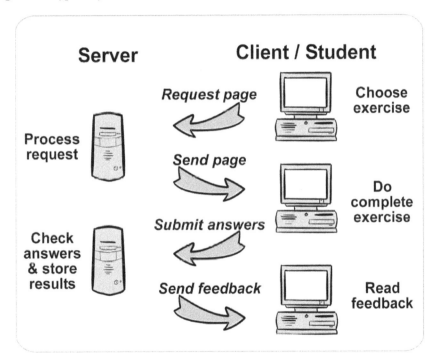

All three of these scenarios go some way towards providing the flexible, enjoyable, interactive experience that client-side exercises can provide, while at the same time adding some measure of access control and performance tracking.

Implications for exercise authors and authoring tools

Any instructor who wishes to create online exercise materials would benefit from clearly understanding the three scenarios above, and the differences in capabilities between the different delivery methods. When planning materials and selecting authoring and hosting tools, it is essential to consider the objective carefully. If the main objective is to do secure testing (insofar as this is possible over the web), and it is necessary to do detailed tracking of student scores, but providing feedback to students as they work on the tests is not important, then the need is clearly for a server-side package. There are two options here. Space can be rented or borrowed on a server running an appropriate package (*Blackboard.com*, for example, offers a hosting service), or the instructor can set up a server-side system herself. In the latter case, some fairly advanced skills (or the help of a competent web server administrator) will be required to install and configure the software. Both of these approaches can be expensive.

On the other hand, if the main objective is to provide self-access learning materials for students doing reinforcement or review work, and there is no real need to fill a database with their scores, then all that is required is an authoring tool that creates client-side exercises (such as *Hot Potatoes*) and an ordinary web hosting account on any web server (such as that provided by an Internet Service Provider, or an institutional web server). Even if the main objective is to run one's own LMS system, some exercise materials will be most suitable for the client-side mode, so it is important to make sure that whichever LMS system is chosen permits uploading and serving of ordinary client-side pages as components of the course material.

The authoring interface itself is also an important consideration. Most LMS systems include online authoring modules that enable instructors to create materials by filling in forms over the web. This can be very convenient, but it does have one limitation: every time the author finishes a question or segment of the exercise, the page will have to communicate with the server to store the data, and retrieve the next input form. This can be frustratingly slow, and can impede the creative flow of building an effective exercise. The alternative is to get a client-side tool that can build the complete exercise on a local machine (and probably test it there too) before the instructor commits to uploading it to the LMS system or the web server. Client-side authoring is generally quicker and more reliable, but involves downloading or purchasing a software package to do it. There is no necessary correlation between the authoring system and the delivery system; *WebCT*, for example, has an online authoring interface, but one can also use a client-side tool called *Respondus* to create *WebCT* quiz materials, which can then be uploaded into the *WebCT* server. Similarly, *Hot Potatoes*, which is a client-side authoring tool, can create output in a format that can be uploaded to *WebCT*, whereas *Swarthmore's Interactive Exercise Makers* are server-side authoring tools that produce client-side exercises. Generally speaking, though, client-side authoring tools can take advantage of the richness of the operating system environment to provide a more sophisticated authoring interface, while web-based server tools are more limited because their interface is constrained by the web environment and the need to communicate between server and client. In addition, if an instructor creates web pages locally and uploads them to a server manually, the final code can be hand-edited in order to customise the appearance or behaviour of the exercise; this is less practical when using a server-based authoring engine.

Security issues

One of the most common concerns raised by instructors who are new to online teaching and learning is the question of security. A teacher in a classroom is used to personal interaction with students, and a high degree of control over those students; if a test is administered, generally there is no possibility that someone other than the student can take the test on their behalf, and all individuals are known to the invigilator. Basic monitoring can prevent serious cheating. Even in a large-scale examination, such as a TOEFL test, students' identities can be verified using ID cards and so on. In the online environment, nothing like this is possible. Communications between client and server can be encrypted to a level that realistically precludes snooping (hence it is relatively safe to use a credit card over the web); in that sense, web interactions can be undertaken with a high degree of security. However, all the systems and protocols involved in such secure transactions are concerned with hiding the transmissions between the client and the server from anyone outside who may try to intercept them. This is the primary objective of the security components in web browsers and servers.

This kind of security is useful in some senses (for example, it can protect privileged communications between instructors and students, and prevent grades and other sensitive information from falling into the hands of outsiders). There are also a number of things that an LMS system can do on the server side to contribute to the security of online testing: the transmission of the test materials may take place following password authentication of the student, and can be time-stamped; the student may be limited to only one access to the materials, and the whole communication may be done via Secure Socket Layers (SSL) using a high level of encryption.

However, this does not help in any way as far as cheating is concerned. Any student who is not directly supervised by an invigilator in the same room has a tremendous range of cheating options available when doing an online test. For example, s/he may summon a collection of experts in the field. S/he may contact other experts live online using chat clients, and interact with them continuously. S/he may have a collection of textbooks open, and search the web for answers as well. In this sense, setting up secure online testing is a completely unsatisfactory endeavour. The only way to do it is a combination of secure technology and direct invigilation. The invigilation may be done by remote centres who provide a service to the testing agent (as in the case of online *TOEFL* testing, where candidates report to a remote centre and are physically invigilated, although the test itself is delivered over the network); but invigilation must take place. This undermines the convenience and cost benefits of online course delivery, because it is clearly expensive and administratively difficult. This is probably why the providers of the *TOEFL* test, the *Educational Testing Service,* are now in the process of closing 84 computer-based testing centres outside North America on the grounds that they are too expensive, and replacing them with paper-based testing centres.

In the light of this, it is hardly worth thinking about online testing as in any way analogous to what is normally done when a test is delivered in a traditional classroom. We will fare far better if we can abandon any fondness we have for traditional testing methods when we enter the online environment. The function of quiz or exercise scoring in online teaching should be primarily diagnostic rather than evaluative; students can use scores to assess their progress, and instructors can use them to diagnose problems with materials (including the quizzes themselves) and teaching points that need further reinforcement. If online exercises are viewed in this way, their main purposes are:

1. Actual instruction (in other words, teaching through discovery learning, etc.). An example of this might be an online action maze which a student has to solve by making a series of correct choices, in the course of which a teaching point is conveyed.
2. Diagnosing areas in which a particular student has problems, and directing him/her as directly as possible to remedial work which can help.
3. Confirming a student's competency or mastery in a particular area.
4. Confirming for the instructor that the class as a whole has mastered or failed to master particular points.
5. Revealing problem areas in the exercise materials themselves, which the instructor or instructional designer can then remedy.

For purposes 1, 2 and 3, scores need not be harvested or stored at all. In fact, it is arguable that *not* storing or transmitting scores (and making this policy clear to the student) makes the exercises more effective, because the student is not afraid to make mistakes and score poorly. Being able to try known wrong answers without penalty (and thus confirm for oneself not only that they *are* wrong, but also *why* they are wrong) is an extremely helpful option, and a student who knows that no tracking is taking place is much more likely to indulge this kind of curiosity. In addition, the online environment is particularly suited to purpose 2, because a good exercise can include helpful feedback which automatically directs the student to extra materials elsewhere on the network. Client-side online exercises (without tracking) are therefore suited to taking the place of some types of classroom instruction. Obviously, though, the quality of the exercise material (and the depth and range of feedback and referral information – see Felix this volume) is crucial if learning is to take place, just as good teaching is essential in a classroom.

Summary recommendations for exercise types

1. Client-side exercises are most effective for instruction, self-assessment, and reinforcement.
2. Server-side exercises with score-tracking are effective for diagnostic purposes, and loose evaluation.
3. For serious evaluation, open-ended (not computer-graded) questions should be used. Responses should be marked manually, and these assignments should be set fairly often throughout a course.
4. A certain amount of cheating must be expected and accepted. This can be minimized through security measures, but never reliably eliminated. With online learning, the student carries more of the responsibility for his/her progress than may be the case in a traditional classroom anyway (see Beaudoin 1996).
5. Really secure testing can only be done by gathering students together for personal invigilation.

LMS standards and standard exercise formats

Having chosen the appropriate type of interactive exercise, the instructor will need to select from a range of authoring tools and hosting environments that promise to help build and deliver the materials. One very important issue to consider when evaluating these tools is the reusability (or otherwise) of the data. When time is spent typing exercise data into a program or database, it is vital that if necessary, the data can be retrieved and used in another location, environment or program. Instructors (or their institutions) can repurpose, sell or exchange the learning materials with another institution, or for another course. If the materials 'describe themselves' in enough detail (see 'metadata' below), then students may be able to find and use them independently, with no need for mediation by an instructor. This table shows the range of degrees to which exercise data might be reusable:

Figure 4. Degrees of data reusability

Reusability	Explanation
None	The data cannot be extracted from the LMS system or authoring tool at all.
Limited to inside the same system	The data can be extracted in a form which is only readable by that LMS system, so it can be moved from one course or zone to another within the same LMS system, but there is no way to recover it for use in other contexts. An example might be a CD-ROM created with a package such as *AuthorWare*, in which exercise data is embedded directly into program code.
Recoverable in proprietory but parsable format	The data can be uploaded and downloaded in file formats which are documented and potentially readable, but do not conform to any recognised standard. *WebCT* version 3.6, for example, allows the import and export of quiz question data in the form of textfiles formatted according to a proprietary, but well-documented, format.
Recoverable in standard language	The data can be downloaded in a language (such as HTML or XML) which is universally readable, but the organisation of the data itself may be proprietary. *Hot Potatoes* data files are like this; they are written in a standard language (XML), but have their own format which is used by very few other programs.
Recoverable in standardised document structure	The data can be downloaded in a format which conforms to a recognised industry standard designed for the sharing of this kind of data between systems (such as the *IMS Consortium QTILite*). Many of the larger commercial LMS systems now claim support for such standards.

However, standards for the structuring of instructional data are in their infancy, and the maze of competing and parallel specifications under development is both confusing and off-putting to most users. The following is a brief introduction to the main standards and their purposes. The standards can be divided into four areas, which we will fully address in turn.

1. Metadata (formats for describing the content of learning material).
2. Packaging (formats for storing and delivering various types of content so that they can be read and used by multiple systems).

3. Sequencing (formats for specifying the order in which materials should be presented, and what requirements might be necessary before a student can proceed to the next section).
4. Learner tracking (formats for identifying and authenticating a learner, and for storing information about his/her progress, achievements and preferences).

Metadata

Metadata is information about information. If one were packing away a lot of old books into boxes, one might write on the outside of each box a brief description of the type of books it contains – hardback or paperback, subject area, and so on. This is what metadata standards are designed to do. Any kind of 'learning object' (a vague, currently faddish term which means any self-contained piece of teaching material, from a short-answer question to a complete textbook) can be described in terms which enable it to be found easily and integrated appropriately into learning sequences. Metadata might include the following sorts of information: *title, authors, publication date, language, institution, subject, level, main concept* and so on. The basic idea is that if metadata is in a standard format, an author can easily search a range of existing resources for learning objects that can be reused in a course under development, rather than creating all the necessary materials from scratch. The metadata is typically included in a document header or in an index database, so that it is not necessary to search through the actual content or instructional material. An example would be The *Co-operative Learnware Object Exchange (CLOE)*, described as 'a collaborative project of eight Ontario universities to develop an innovative infrastructure for joint development of multimedia-rich learning resources'.

Whatever the aim of the instructional designer, and whatever authoring tool is being used, including some descriptive metadata is undoubtedly good practice. The important thing is to strike a balance between adding enough data to make the material usefully self-describing, and investing too much time in detailed analysis and description of material that may not have a long projected lifetime, or may not be of much interest or use to anyone else. A simple and effective system is proposed in the *Dublin Core* standard, which has only a handful of basic metadata elements. Here is an example of *Dublin Core* metadata encoded in XML. The following code describes our chapter of this book:

```
<?xml version="1.0"?>
<rdf:RDF xmlns:rdf="http://www.w3.org/1999/02/22-rdf-syntax-
ns#" xmlns:dc="http://purl.org/dc/elements/1.1/">
<rdf:Description rdf:about="">
<dc:title>Servers, clients, testing and teaching</dc:title>
<dc:subject>Online instructional exercises</dc:subject>
<dc:creator>Stewart Arneil</dc:creator>
<dc:creator>Martin Holmes</dc:creator>
<dc:contributor>Half-Baked Software, Inc.</dc:contributor>
<dc:language>en</dc:language>
<dc:description>This paper deals with the differences between
creating instructional exercises for simple clients and
complex server-side systems, and with emerging industry
standards for e-learning.</dc:description>
<dc:date>2002-05-31</dc:date>
```

```
<dc:coverage>Emerging norms and standards in 2002</
dc:coverage>
<dc:type>Text document with graphics</dc:type>
<dc:relation>Is part of Language Learning Online: Towards Best
Practice (ed. Uschi Felix)</dc:relation>
<dc:source>Language Learning Online: Towards Best Practice
(ed. Uschi Felix)</dc:source>
<dc:publisher>Swets & Zeitlinger</dc:publisher>
<dc:rights>Restricted</dc:rights>
<dc:format>text/xml</dc:format>
<dc:identifier>[ISBN number not yet assigned]</dc:identfier>
</rdf:Description>
</rdf:RDF>
```

An authoring tool should be able to aid the designer in adding appropriate metadata for materials in a standardised format, such as this, but without the need to understand the code itself.

Packaging

Packaging standards define how actual learning materials are stored and transmitted. Perhaps the most significant is the IMS standard, under development by the *IMS Global Learning Consortium*. Over the past few years, this consortium has been developing a set of standard data structures (implemented in XML) for storing instructional data such as questions, answers, feedback etc. (The *IMS* standards also cover metadata, sequencing, learner tracking data, and a range of other information.) The core specifications were released in version 1.2 in early 2002. *IMS* specifications are intended to allow both instructional material and student performance information to be shared between different LMS systems and authoring tools. For the purposes of this paper, which is concerned mainly with interactive exercise materials, the most important component is the Question & Test Interoperability specification (*QTI*, also available in a pared-down form as *QTI Lite*). However, the current *QTI* specification, even in its 'Lite' format, is remarkably complex. Even a simple multiple-choice question encoded in this format becomes a long and complex document.[1] This is probably a result of the need for a global specification like *QTI* to support such diverse user groups as corporate training departments, university courses, technical certification testing, and so on. It is interesting to note that the *Dublin Core* group is working hard against this tendency for standards to become too complex and unwieldy to be practical:

> The Dublin Core element set has been kept as small and simple as possible to allow a non-specialist to create simple descriptive records for information resources easily and inexpensively, while providing for effective retrieval of those resources in the networked environment. Hillmann (2000:1)

The designers of *QTI* seem to be taking the opposite approach. Nevertheless, despite the sophistication of the *QTI* data specification, it does not seem capable of encoding even the small range of interaction types possible with *Hot Potatoes*. For example, *Hot*

---

1.   See example at end of this chapter.

Potatoes allows the insertion of any HTML data (including links, pictures, media files, embedded objects etc.) inside any element of a multiple-choice question, so that one could include two pictures in the question, a link to a sound file in each of the multiple-choice answers, and present responses in the form of videos. The *QTI Lite* specification appears to disallow (or, at least, does not clearly allow) this kind of mixed data. A typical *QTI Lite* question would be one (and only one) of [text, picture, sound]. The ability to compose sophisticated HTML blocks for use as questions, answers or feedback, and include any embedded elements one chooses, seems to go against the grain as far as the specification is concerned. It is likely that a range of basic interaction types will be accommodated in the standard, and that materials based on these types of exercise will be encodable at a detailed level; on the other hand, other more unusual, idiosyncratic, or specialised interaction types will have to be packaged in a less granular way, as single objects. For instance, a standard multiple-choice exercise can be encoded using the *QTI* system such that any *IMS*-compliant LMS can deliver it, and retrieve a student's results on a per-question/per-answer basis. However, a crossword exercise, which is not explicitly accommodated by the *IMS* data structures, would have to be packaged as a unit (perhaps as a single interactive web page), and simply described in metadata terms on the basis of content, level, objectives, and so on. The crossword could be stored, referenced and delivered by the LMS system – and in fact a basic score might even be retrieved and stored – but the LMS system would not 'know anything about' the individual clues and answers in the crossword.

Most packaging systems currently under development use XML, and as explained below, XML can easily be transformed from one structure into another. Therefore, as long as XML is the packaging method, the standard does not matter too much to the average content creator. However, if an authoring tool is highly constrained by the need to conform to a specification, then its utility may be limited.

Sequencing

Both the *IEEE Learning Technology Standards Committee* and the *IMS Consortium* have working groups on sequencing, but no specification documents have yet emerged from either; according to SCORM (the Sharable Content Object Reference Model proposed by the *Advanced Distributed Learning* organisation), the *IMS* is at the time of writing about to release a 'simple sequencing' specification. It looks as though the standards will not only address simple series arrangements, but also allow the creation of multiple paths through materials. This would enable the LMS system to respond 'intelligently' to student performance, fast-tracking quick learners while also providing remedial or reinforcement material for those who are struggling. Sophisticated sequencing is useful in creating customised courses for individual learners or groups of learners, including only what they need, and excluding material which is below their level, already covered elsewhere, or irrelevant to them (see Capuano et al 2000).

Learner tracking

Learner tracking data comprises a range of information, including:
- The learner's identity
- Learner preferences (governing, for example, how the material should be presented)
- The date and time a piece of material was accessed by the learner
- How long the learner spent working on the material
- Learner responses (both right, wrong and unmarkable)
- Scoring information resulting from the interaction
- Learner comments on the material
- Outcomes of the interaction (for example, what a learner is now entitled to access as a result of successfully completing an interaction)

The *IEEE Learning Technology Standards Committee specification 1484.11.1* (section 7) includes several sections on learner tracking, covering all of the above. The *IMS QTI* specification includes detailed data structures for reporting of results (*IMS Question & Test Interoperability: Results Reporting Information Model*); in the *IMS* model, the reporting system forms part of the question/test structure itself. The advantage in standardising this type of information is that learners often move from one educational environment to another, and given universal data standards, their records can easily move with them. Also, giving or restricting access to learning materials in other systems based on a learner's competence level is more practical. (All of this, of course, depends on the existence of an effective articulation system between institutions, a far thornier problem than that of data encoding standards.)

The issue of learner tracking is more administrative than pedagogical, but nonetheless, it may be necessary to make decisions about the type of information recorded and stored when students access materials, especially when authoring those materials within a large *IMS*-compliant LMS system.

Disadvantages of standards

Generally speaking, web and software developers (such as the authors) tend to be strongly in favour of universal standards for data storage and transmission. However, in this case, there are one or two factors that might cause us to be a little cautious with regard to standards such as those under development by the *IMS Global Consortium*.

First, the idea of learning objects presupposes that we wish to make our materials available for sale, reuse, adaptation, and so on. That may not be the case; we may not want to make it easy for our employers, institutions, colleagues or competitors to distribute our work widely, for profit or otherwise. As Noble has pointed out:

> *Once faculty put their course material online, moreover, the knowledge and course design skill embodied in that material is taken out of their possession, transferred to the machinery and placed in the hands of the administration. The administration is now in a position to hire less skilled, and hence cheaper, workers to deliver the technologically prepackaged course. It also allows the administration, which claims ownership of this commodity, to peddle the*

course elsewhere without the original designer's involvement or even knowledge, much less financial interest. The buyers of this packaged commodity, meanwhile, other academic institutions, are able thereby to contract out, and hence outsource, the work of their own employees and thus reduce their reliance upon their in-house teaching staff. Noble (1998:1)

Secondly, the authoring of learning materials is undoubtedly a creative, imaginative process, and a degree of experimentation is healthy if new types of material are to evolve and gain currency. The institutional and industrial drive to standardise materials so that they fit easily into the current paradigm, as described in one or another learning object specification, can only restrict creativity to some degree. There is surely a danger that learning materials may become progressively less various, distinctive, and innovative, as instructors and instructional designers work increasingly within the confines of structured LMS systems. In addition, preparing or adapting materials for an LMS using a complex tagging system involves substantial overhead in terms of developer time; this may not be worthwhile in the case of materials whose lifetime is not expected to be long (for example, topical language-teaching materials based on current events). The cost of adding all this information is particularly high for small-scale developers with a very specific audience in mind, such as teachers developing materials aimed at specific groups of students or even individuals.

A standardised data structure may in itself constitute a limitation on the range and type of materials that can be created, so sophistication and flexibility in materials design may be compromised by the need to conform to the requirements of the standard. For instance, a *Hot Potatoes* Crossword exercise cannot be encoded using the *IMS* standards, because crossword puzzles do not constitute a recognised component of the *IMS* specification. However, *Hot Potatoes* itself saves the crossword file in an XML structure which can be read by any XML parser. Therefore, someone with the requisite skills could, for example, write an XSLT stylesheet which would render the crossword in a useful or attractive fashion despite the fact that the actual data structure in the *Hot Potatoes* file is 'proprietary'. As long as the encoding language used (in this case XML) is standard, then the data remains accessible, albeit with some effort. If a future *IMS* standard were to add support for crossword puzzles, it would be relatively easy to write another XSLT converter to turn the *Hot Potatoes* data file into a block of *IMS*-compliant code. The point here is that use of a *QTI*-based authoring tool would preclude the creation of a crossword exercise, so compliance here constitutes a limitation.

Descriptive frameworks based on the notion of learning objects also embody a particular pedagogical approach – a highly analytical one, in which learning materials are viewed as fundamentally modular. The effectiveness of this modular approach depends on consistent levels of 'granularity' (Porter 2001:49). Materials which are more holistic in conception fit less easily into the structure. The overall aim of the learning objects movement is to enable mix-and-match course composition on-the-fly, by instructors, institutions, or even students themselves; while this may be appropriate for some subject areas and training systems, it is by no means clear that the academic community, particularly in the humanities and fine arts, would want to accept such an approach to education.

In addition, we are not convinced that the aims and requirements of the diverse group of interested parties developing the standards are actually consonant with each other. For example, surely the needs and interests of a company such as Microsoft (who provide a resource kit for educators through their *Learning Resource Interchange* initiative) will be largely dissimilar from those of a university professor teaching a course on the ethics and morality of capitalism. There is a clear risk that the needs of corporate training departments, defence institutions, and the commercial providers of e-learning tools (all of whom are wealthy in comparison with the state education sector) will tend to dominate the design and implementation of the specifications, and therefore ultimately the nature of the materials created within them. The *IMS QTI* specification overview states:

> *The QTI specification, like all IMS specifications, does not limit product designs by specifying user interfaces, pedagogical paradigms, or establishing technology or policies that constrain innovation, interoperability, or reuse. (http://www.imsglobal.org/question/qtiv1p2/ imsqti_res_infov1p2.html#1403937)*

However, most people will be authoring their standards-compliant materials using authoring tools created by members of the consortium or companies dependent on them, because the data structures are so complex that tools are required. This means that what can be done will be constrained by what the tools allow, so the consortium members may have more effective control than even they intend over the nature of the material which is authored for, and stored in, standards-compliant LMS systems.

Finally (and perhaps in mitigation), there is a big difference between creating complex standards and actually implementing them. While the standards themselves continue to grow in size and complexity, it becomes more and more difficult for any given LMS system or authoring package to implement them in their entirety. If one LMS implements 50% of the standard, and another implements a different 50%, then true interoperability cannot be achieved; moving data from one system to another will inevitably result in the loss of some information. The *IMS QTI* listserv hosts frequent complaints from those attempting to implement standards, to the effect that no existing implementation even approaches completeness, and that current flaws in the standards would actually make it impossible to implement them fully. The *SCORM* specification itself explicitly allows an LMS to be conformant [sic] even in the case of incorrect implementation:

> *If the LMS **incorrectly** implements one or more SCORM Version 1.2 Run-Time Environment Data Model Optional Elements, and does not implement any other optional data model elements correctly, then the LMS is still considered to be LMS-RTE1 conformant as long as the criteria above are met. (SCORM version 1.2 Run-Time Environment Conformant Minumum standard, Advanced Distributed Learning 2002)*

In sum, the drive to standardise the structure and description of learning materials undoubtedly has administrative benefits at the institutional or corporate level, and does make it easier to find and reuse materials in a very efficient (and profitable) way; however, it works in opposition to the decentralising and liberating features of web-based technology which have been so stimulating to educational innovation in the last few years. As we point out below, though, the very nature of XML makes it easy to transform data from one format into another, so provided that the data is in some kind of well-structured XML, it can always be reformatted to comply with one specification or another.

Recommendations regarding LMS standards

1. Compliance with currently-developing standards (such as *SCORM* or *IMS* specifications) is NOT essential at the present time. These are moving targets, and the claim of compliance by one tool or service-provider does not at the current time ensure that data can be easily ported to another, because implementations of the standards are at best partial.
2. Using a standard data encoding is essential. The only real choice here is XML. Provided the data is encoded in XML, it can be transformed into any other standard format.
3. Materials should always include metadata, if possible in a standard format such as the XML implementation of *Dublin Core*. This will make the materials easier to find, use, and catalogue, as well as asserting the author's intellectual property and authorship rights in a formal recognised manner.

Conclusions

Authoring tools and Learning Management Systems have proliferated in the last few years, and the choices facing instructors and instructional designers can be daunting. In order to evaluate potential tools, it is essential to have at least a basic understanding of how the underlying technologies for authoring and delivering online content actually work, and the strengths and weaknesses of the various options. In particular, server-side and client-side models have very different capabilities and limitations, both for authoring and delivery. Even where one is forced into the use of a particular tool or LMS by institutional pressure, it is often possible to circumvent some of its limitations by incorporating other types of material (for example, by delivering client-side materials from a server-side LMS). For any given set of teaching requirements, a variety of tools of different types are likely to be needed in order to allow a sufficiently wide range of interaction types and tracking capabilities (not to mention creative freedom on the part of the designer).

Where evaluation is required, a sophisticated understanding of the various types and levels of security is essential; the protection of data is not the same as the prevention of cheating, and formal evaluations, such as final tests, should probably not be undertaken online unless strict physical invigilation is possible. A good alternative strategy is the use of frequent open-ended assignments such as short essays marked or graded by an instructor, so that, although cheating is possible, it requires a lot more effort.

The dominant model for instructional materials throughout the e-learning industry is the learning object. Learning objects, by their very nature, are sharable and reusable. Not all types of material fit easily into this model (highly topical materials, for example, may be completely transient), and not all materials creators may wish to facilitate this kind of exploitation of their work. The learning object model is primarily driven by large institutions and corporations seeking cost benefits through re-use of data, and this has also contributed to the emergence of standards for interoperability of LMS and authoring systems. This is undoubtedly a good thing; however, the major specifications are still very much in their infancy, and poorly implemented across the industry, so not much is to be expected in terms of true interoperability for some time yet. For example, a claim of 'SCORM-compliance' on the part of a vendor is promising in the sense that it demonstrates awareness of, and respect for, industry standards, but it may not constitute a useful feature from the end-user's point of view. In addition, the need to conform to standards and specifications can limit the creativity of a designer or the capabilities of an authoring tool.

In view of this, well-structured, well-described data is the surest guarantee of long-term portability and usefulness, and XML provides the most viable format because of its broad support both inside and outside the e-learning community, and its solid Unicode-based language support. In XSLT, XML has a companion tool for transforming data from one format to another, thus ensuring compatibility with future standards. Metadata should also be included in a recognised format (such as *Dublin Core*), to enable indexing, searchability and clarification of intellectual property claims.

References

Advanced distributed learning (2002). scorm version 1.2 conformance requirements version 1.2 http://www.adlnet.org/index.cfm?fuseaction=rcdetails&libid=285

Beaudoin, M. (1996). The Instructor's Changing Role in Distance Education. *The American Journal of Distance Education, 4, 21-29.*

Capuano, N., Marsella, M & Salerno, S. (2000). ABITS: An Agent Based Intelligent Tutoring System for Distance Learning. Proceedings of the International Workshop on Adaptive and Intelligent Web-based Educational Systems held in Conjunction with ITS 2000 Montreal, Canada http://littlebottom.cl-ki.uni-osnabrueck.de/its-2000/final.pdf

Heift, T. (2001). Error-specific and individualised feedback in a Web-based language tutoring system: Do they read it? *ReCALL, 13, 99-109.*

Hillmann, D. (2000). Using Dublin Core, Dublin Core Metadata Initiative working draft http://www.dublincore.org/documents/2000/07/16/usageguide/

Noble, D. (1998). Digital Diploma Mills: The Automation of Higher Education. *First Monday, 3* http://www.firstmonday.dk/issues/issue3_1/noble/index.html

Porter, D. (2001). Object Lessons from the Web: Implications for Instructional Development. In G. Farrell, (Ed.), *The Changing Faces of Virtual Education,* 47-69. The Commonwealth of Learning.

Pujolà, J-T. (2001). Did CALL feedback feed back? Researching learners' use of feedback. *ReCALL, 13, 79-98.*

ReadyGo WCB: 1-0-379 (2001). An Introduction to AICC, SCORM, and IMS (XML)
    http://www.readygo.com/aicc/.
Robson, R. (2001). Learning Technology Standards: Status and Direction
    (PowerPoint presentation) http://www.eduworks.com/Web/Docs/
    Status_and_Direction.ppt
World Wide Web Consortium (1999). Resource Description Framework (RDF)
    Model and Syntax Specification: W3C Recommendation 22 February 1999
    http://www.w3.org/TR/1999/REC-rdf-syntax-19990222/

Websites

Advanced Distributed Learning – http://www.adlnet.org/

AuthorWare – http://www.macromedia.com/software/authorware/

Blackboard – http://www.blackboard.com/

Canvas Author Wizard – http://www.canvaslearning.com/

Co-operative Learnware Object Exchange (CLOE) –
    http://lt3.uwaterloo.ca/CLOE/overview.html

Dublin Core – http://dublincore.org/documents/1999/07/02/dces/

Educational Testing Service – http://www.ets.org/icenter/centers.html

FormMail.pl – http://www.scriptarchive.com/formmail.html

French Revision – http://www.frenchrevision.co.uk/index.htm

Hot Potatoes – http://web.uvic.ca/hrd/hotpot/

hotpotatoes.net – http://www.hotpotatoes.net/

IEEE Learning Technology Standards Committee – http://ltsc.ieee.org/

IEEE Learning Technology Standards Committee specification 1484.11.1 –
    http://ltsc.ieee.org/doc/wg11/1484_11_1_D6.doc

IMS Global Learning Consortium Inc. – http://www.imsglobal.org/

IMS Question & Test Interoperability: Results Reporting Information Model –
    http://www.imsglobal.org/question/qtiv1p2/imsqti_res_infov1p2.html

LUVIT – http://www.luvit.com/

Microsoft Learning Resource Interchange –
    http://www.microsoft.com/elearn/support.asp

QTI – http://www.imsglobal.org/question/index.html

QTI Lite – http://www.imsproject.org/question/qtiv1p2/imsqti_litev1p2.html

Quandary – http://www.halfbakedsoftware.com/quandary/

QuestionMark – http://www.questionmark.com/uk/home.htm

Quia – http://www.quia.com/

Respondus – http://www.respondus.com/

Swarthmore Interactive Exercise Makers – http://lang.swarthmore.edu/makers/

TexToys – http://www.cict.co.uk/textoys/

TOEFL – http://www.toefl.com/

WebCT – http://www.webct.com/

All websites cited in this chapter were verified on 09.09.2002.

Example of a single m/c question with four answers, encoded according to QTI Lite using the Canvas Author Wizard:

```xml
<?xml version="1.0" encoding="UTF-8"?>
<!DOCTYPE questioninterop>
  <questioninterop>
   <item title="md ident="item">
     <presentation label="can_presentation">
       <material label="can_question_material">
         <attext type="text/plain"><![CDATA[What does the
acronym QTI stand for?]]></mattext>
       </material>
       <response_lid ident="question" rcardinality="single">
         <render_choice shuffle="yes">
          <flow_label>
            <material>
             <mattext type="text/
plain"label="bullet"><![CDATA[*]]></mattext>
            </material>
            <response_label ident="a">
             <material>
               <mattext type="text/plain"><![CDATA[Quick Test
Initiator]]></mattext>
             </material>
            </response_label>
          </flow_label>
          <flow_label>
            <material>
             <mattext type="text/
plain"label="bullet"><![CDATA[*]]></mattext>
            </material>
            <response_label ident="b">
             <material>
               <mattext type="text/plain"><![CDATA[Question and
Text Interoperability]]></mattext>
             </material>
            </response_label>
          </flow_label>
          <flow_label>
            <material>
             <mattext type="text/
plain"label="bullet"><![CDATA[*]]></mattext>
            </material>
            <response_label ident="c">
             <material>
             <mattext type="text/plain"><![CDATA[Quiz Template
Installer]]></mattext>
             </material>
            </response_label>
          </flow_label>
          <flow_label>
            <material>
             <mattext type="text/
plain"label="bullet"><![CDATA[*]]></mattext>
            </material>
            <response_label ident="d">
             <material>
               <mattext type="text/plain"><![CDATA[Quotidian
Terminological Instantiator]]></mattext>
```

```
          </material>
            </response_label>
          </flow_label>
          </render_choice>
        </response_lid>
      </presentation>
      <resprocessing>
        <outcomes>
          <decvar varname="score" vartype="integer"
defaultval="0"></decvar>
        </outcomes>
        <respcondition title="AllCorrect">
          <conditionvar>
            <varequal respident="question">d<varequal>
          </conditionvar>
          <displayfeedback linkrefid="displayCorrect"></
displayfeedback>
        </respcondition>
        <respcondition title="allWrong">
          <conditionvar>
            <not>
              <varequal respident="question">d</varequal>
            </not>
          </conditionvar>
          <displayfeedback linkrefid="displayWrong"></
displayfeedback>
        </respcondition>
        <respcondition title="adjustscore">
          <conditionvar>
            <varequal respident="question">b</varequal>
          </conditionvar>
          <setvar varname="score" action="Add">100</setvar>
        </respcondition>
      </resprocessing>
      <itemfeedback ident="displayWrong">
        <flow_mat>
          <material>
            <mattext type="text/plain"><![CDATA[Sorry! Try
again.]]></mattext>
          </material>
        </flow_mat>
      </itemfeedback>
      <itemfeedback ident="displayCorrect">
        <flow_mat>
          <material>
            <mattext type="text/plain"><![CDATA[Well done! The
Question and Test Interoperability standard is being developed
by the IMS Consortium.]]></mattext>
          </material>
        </flow_mat>
      </itemfeedback>
    </item>
  </questioninterop>
```

5

Engaging the learner – how to author for best feedback.

Paul Bangs, Consultant in Language Technology, U.K.

Introduction

If one had to prioritise reasons for using a computer to assist with language learning, then arguably the role of feedback would be high on anyone's list. Feedback is, in many ways, the key to what a computer does best, and can be the link between the technical power of the program and the pedagogical needs of designer and user. Unfortunately, there have been relatively few serious evaluations of the use and effectiveness of feedback in language learning, and one is often left with anecdotal evidence or common sense as a criterion with which to judge effective design. This is also addressed by Felix in this volume.

It is not the intention of this chapter to go further into the field of cognitive research into feedback types. Much good work has been done and is continuing in this field, (Pujolà 2001, Heift 2001, Heift 2002). What appears to us more interesting is to attempt to examine why good feedback is lacking in so many computer-based programs seen today, especially those available via the Internet.

After many years of experience, experimentation and research (although the last is very scarce), the use of feedback should, by now, have been presenting the CALL community with well-established patterns and examples of best practice. However one has but to carry out a cursory Internet search to see that in huge numbers of cases the application of feedback techniques is either poor or lacking. Why this should be is the subject of much speculation. But the reasons can, in our opinion, be investigated.

This chapter will begin by examining the various types and classifications of feedback and their role in the learning process. It will go on to consider the ways in which multimedia could be used to better effect in the feedback process, and offer suggestions for more complex feedback structures than the standard elements usually seen. After consideration of the constraints which using the Internet might place on these systems, attention will then be focused on the ways in which effective authoring can be used to create the most effective routines. The chapter will conclude with a discussion of the *MALTED* authoring tool, which offers the potential to empower better use of pedagogically appropriate feedback systems.

Why feedback?

Feedback can be the feature which enables independent learning. It can also be the key to individualised learning if handled well. Both these factors can in turn influence more strategic and methodological issues, such as classroom design, curricular planning and more. But all of that will count for nothing if the design of the feedback which the learner receives is not well planned and pedagogically sound. Poorly designed feedback is far too common, and this is especially so when one reviews many programs offered on the Internet. The reasons for this are many, and need to be examined in more detail.

What is feedback?

There have been many attempts to define feedback over the years. Skinner (1958) described feedback in terms of any information that follows a response and allows a learner to evaluate the adequacy of the response itself. After more than 60 years this is still not a bad definition. Others (Bationo 1992) have widened the definition to include an event as well as a response, which is also significant, and indicative of the advances made in the use of multimedia in the intervening years. In this context, it is not altogether surprising that earlier studies and advice, in the years before even sound cards were available in PCs, concentrated on error correction for text-based exercises, as in Pusack (1983). Whilst Skinner may have viewed feedback in the context of reinforcement and programmed learning, nevertheless the early efforts had one positive outcome which designers would do well to bear in mind today, and that is the need for carefully elaborated instructional design (see Hémard this volume), a theme which will be revisited later in this work.

Types of feedback

Buscemi (1996) makes useful distinctions of types of feedback. The basic types are 'Knowledge of response', where only an assessment is given to the learner as to whether they are right or wrong, 'Knowledge of correct response' where the right answer is given, and 'Answer until correct' which offers the learner the chance to give a correct response. This form of feedback may be described as 'verification' feedback. What is perhaps more interesting is the use of 'elaborative' feedback which attempts to give the user more information about their response. Within this category, Buscemi distinguishes between 'adapted' and 'adaptive' forms. Whilst this may be a somewhat simplistic approach, it is still useful to distinguish between the use of 'adapted' feedback which is inherently generic and universal, and 'adaptive' systems which can be dynamic and adjustable for different classes of end user.

Alongside the above interpretations, another form of distinction can be made, and one which appears to offer more practical use. This is the distinction between 'extrinsic' and 'intrinsic' feedback. The former has been described as 'a meta-level comment, external to the action the student makes' (Laurillard 1993:36-38). Thus an external comment, rating (score) or sound, obviously offer limited specific help in the learning process. But going even further than this, it can be stated that the provision of information needed is often in itself an extrinsic intrusion into the communicative process. In 'real' language situations, learners are rarely told "Well done!" or "Try again". They encounter feedback, usually without realising it, when their efforts are greeted with a shrug of incomprehension, the request to repeat the utterance, or the right (or wrong) outcome of the transaction or request. This is the ideal type of feedback, and one which enables the learner more rapid acquisition of language patterns. This is what is meant by 'intrinsic' feedback, and seems to us to represent a more satisfactory match to constructivist learning approaches (see also Orellana et al 2001). To reproduce intrinsic feedback patterns in computer programs is not always easy, but that should surely not be a bar to attempting to achieve the ideal.

A major point to be considered is one that arises from attitudes beyond the use of computers and technology. No doubt in part due to the 'cumulative-progressive' nature of language learning, one has, over the years, often seen examples of the over-use of testing, ostensibly to assist the progress of the learner. This has survived the demise of the audio-lingual approach, and is used much less in the classroom these days, but seems in turn to have found a home in many computer routines. It is important to distinguish between practice and testing. Indeed, it can be argued that one should be more careful in making the distinction between various phases of learning – exposition, activity, exercise and test. The distinction is not always completely valid, as there are many different learning situations, but our experience has shown that failure to understand the difference between these phases and to link them into a coherent sequence of learning almost always results in impoverished routines. The subject of this chapter precludes further study of the curricular issues, but what is of interest here is the role of feedback in the various learning phases.

Another way of expressing the problem is to state the difference in terms of theoretical questions which a learner might ask:
- "What do I need to know?"
- "What did I do wrong?"
- "Why did I get it right?"
- "How am I doing?"

To this one can add a further element which is not always transparent to the user, which is: "What happens next if I make this decision?"

The "What do I need to know?" question may not always be recognised as feedback, but in fact the standard use of hyperlinks is almost by definition an example of feedback given on the request of the end user. But this is not what is of primary interest for the current study. More obviously seen as feedback issues are the questions relating to the reasons why a learner performed well or not – and at this stage it should be remembered that learners need to be informed why they got something correct, just as they will benefit from knowing why they made an error. This is often overlooked when designing materials.

There is a distinction to be made between different types of feedback, which can usefully be divided into the headings of *intrinsic* and *extrinsic*. Language learning exists for learners to be taught how to act in real situations in the target language and country. In 'real' usage of language, we are continually in receipt of feedback, without realising it. Every time we see our interlocutor nod their head in agreement (or otherwise) we are being reinforced by a type of feedback that is most definitely *intrinsic*. On a simple transactional level, if we use the target language to order a drink in a bar and it arrives, the very act that it is received is intrinsic feedback. Receiving something which was not asked for is also feedback. A (true life) example is requesting a *manzanilla* in a Spanish bar and receiving a camomile tea, when what was desired was another *manzanilla* – a type of sherry. Although this may seem trite, the implications are clear. The more one can match the language learning to the real situation, the more effective the learning may be. Thus intrinsic feedback is an ideal to be aimed at wherever possible.

Extrinsic feedback, on the other hand, arises from 'outside' the context. It may be equated to a 'teacher response' to the input from the learner. There is obviously a place for this in many occasions where intrinsic feedback is not possible or not appropriate. However, within this distinction, there are many grades of feedback types, not all equally valid. The very common type of feedback which is often seen is the 'noise' – and we are all familiar with the 'boings', 'cheers' and so on which accompany evaluation of a learner response. They will of course (subject to being culture bound – and there appears to be little if any research done in this area) let the user know that s/he has got an answer correct or otherwise. But that is all. There is no further quantitative information, and no qualitative advice given (see Felix this volume).

The next stage in considering extrinsic feedback is the production of a score. For many, the immediate response to the question of "What is feedback?" tends to relate to scoring – or "How am I doing?". Testing mechanisms which give the learner a score are of course very common, and have their place. We would argue, however, that inappropriate use of scoring feedback is all too common, especially with Internet-based programs. In such cases, the use of scoring may be said to offer more to the teacher than the learner. As a learner, one has to consider the value of being told that one has scored 60% as a result of a series of tests (which are often wrongly called 'exercises'). In the standard educational world of examining, this could be considered a good score, whereas in fact the student has failed in significant amounts to achieve the rating required. This also speaks volumes for poor matching of the learner to the test material. What is truly lacking in the scoring type of feedback is *qualitative* feedback which will help the learner improve performance. Some will argue that the very act of testing, by informing the learner as to the extent of their progress, is useful feedback. However, there is little evidence to support this, and some which contradicts it (Buscemi 1996).

Going beyond the simple scoring feedback mechanism, far and away the most common type to be seen falls into the category of "Well done!" or "Sorry, try again". Once more, instances where this is appropriate and useful will occur. However, it is used far too often as a mechanism, and one needs to explore the reason why this should be so. In theory, if one can write the feedback line "Well done" then one should as equally be able to write a more appropriate line, such as "Yes, that is the phrase we use when saying 'hello' in the morning", or "No, what you chose was the phrase used to greet someone in the afternoon. Listen and try again." (see Pujolà 2001, Heift 2001). Obviously we are now in the realm of qualitative extrinsic feedback. Whilst intrinsic feedback is always to be aimed at (and this will be returned to later), the next best thing is to use qualitative extrinsic feedback which will form part of the learning, rather than the testing process, and will be motivating and constructive for the learner. The reason this is not seen with sufficient frequency is probably due to several factors. First, the actual instructional design of learning routines is often passed on to the programmer, rather than maintained by those in charge of the pedagogy. To digress slightly, it seems clear that more widespread training in instructional design (as opposed to the higher level pedagogical design) in language learning, and a greater dissemination of best practice, is an urgent need. Another issue is that of cost: writing, (and possibly recording) feedback lines for a large piece of courseware implies significant costs which may not be feasible within budget constraints – hence the need to factor in such items when planning the design in the first place. And finally, there may be the constraints imposed by some authoring systems, which do not easily make it possible for designers and content providers to include this type of feedback.

If the qualitative type of feedback is of value, then it is axiomatic that the closer the feedback can match the learner's performance and profile, the more useful it will be. The ultimate aim should surely be to see feedback along the lines of "Sorry, James, but you are still making the same mistake with tenses. Remember what we did in class last week? You should revise from chapter X and see or email me if you are still unsure." The possibilities of such a targeted feedback clearly depend almost entirely on the size of the audience for whom one is writing the routines, and is rarely achieved (though Heift 2001, comes close). But if the software used gives the ability to write qualitative feedback lines, then similar, if not so specific, targeting is easier than might be supposed. Much depends on the authoring system being used, and this will be examined later.

Feedback and multimedia

Alongside the question of what type of feedback should be given, is the complementary one of "Feedback to what?". It is not appropriate here to examine every type of language learning routine, but an example could be taken with multiple-choice routines, probably the most common form of activity seen in computer based learning of all disciplines. Ignoring for the moment the common failure to distinguish between an exercise and a test – we are all familiar with the 'classic' example of a question, three or four options with 'radio' buttons, only one correct answer, and nothing but a score given – one can consider a multiple choice question (MCQ) in a wider context, and this includes True/False activities, which are merely an example of a 'reduced' MCQ activity. Arneil and Holmes also deal with this in the previous chapter. Especially in the field of language, the correct response in real life is rarely a unique one – language is complex and vital, and the multiplicity of manners of expression is one factor which makes computer-based approaches so difficult to design in an appropriate way from a pedagogical point of view. So it is difficult to understand the prevalence of multiple-choice routines where there is only ever one correct answer. Perhaps this stems from the behaviourist approaches prevalent some decades ago. But there is no real technical, and certainly no pedagogical reason why more than one, or even all the options given should not be correct. (One could just add here that to make an exercise with none of the options correct might well serve to de-motivate the learner, who may feel somewhat deceived by this approach!) Once the learner is aware that more than one answer may be correct, one sees a motivational effect added to the value of the task, fully in line with a constructivist approach to learning. If carefully designed, this can also become the basis for task- or problem-based learning activities.

The next point to be considered is the use of multimedia. Again it is unclear as to why feedback routines nearly always tend to be text-based. Experience with some publishers has shown that cost is indeed a major factor, and one can add to this the technical expertise needed to create multimedia resources and the time needed for this process. Whatever the reasons, MCQ routines themselves are usually in the form of text, even where it would be more appropriate to use other media. It does not take a very deep analysis of the possibilities available to realise that the number of combinations of media is very large. Here it is not only feedback that is referred to. In any MCQ routine there are three elements – the stimulus, the options and the feedback. Given that each of these could be either text, graphic or audio in nature (leaving aside video for the moment, as well as multiple combinations), then clearly it is possible to use, say, a graphical stimulus, a number of textual options, to be received by another graphical feedback, and so on, reaching well over twenty combinations. There is no reason, therefore, that a carefully crafted exercise should not tailor the media to the content that is being developed, practised or tested. At the very least, feedback should be appropriate to the specific content – if the choice is to discriminate between several aural options of appropriate phrases, then the feedback should clearly be in an audio format, or at least have this as an option on demand. Naturally, there are implications here for design and production workloads and costs. But this should be seen as an essential item to be factored into the original estimates, rather than an optional extra if there is money left over, which often seems to be the case. Now that the Internet is offering ever faster means of access with broader bands of communication, the use of media other than text becomes even more attractive. And it does not take a great leap of imagination to see how similar techniques for feedback can be applied to other activity types. More information on these points can be found in Felix (1998) and Felix (2001).

Although the MCQ activity seems straightforward – and in most cases is – there are some other possibilities which can be taken into account in the design when varying media are employed. For instance, it is usual to offer a stimulus and a range of options (Figure 1, overleaf).

But instead of the learner having to click on one of the options immediately, these options might take the form of 'mini-applications' themselves, which the learner is invited to view before making the appropriate choice, as represented in the flowchart in Figure 2 (overleaf). A similar approach was used by Pujolà (2001:83-84).

If carefully designed, such a system can offer a much richer learning experience, since the variety of optional responses given can in themselves represent learning opportunities. Now, to add to this a rather different form of feedback, in which the learner is shown the consequences of their choice *before* (or instead of) being told whether they have chosen wisely or poorly, and one can see that this is approaching the desirable outcome of an ***intrinsic*** feedback.

Figure 1. A 'standard' MCQ exercise

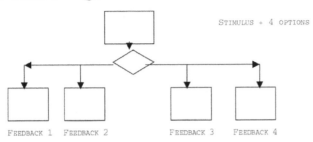

Figure 2. A 'preview' MCQ exercise

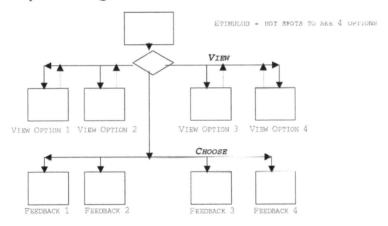

Figure 3. A 'branching' MCQ exercise

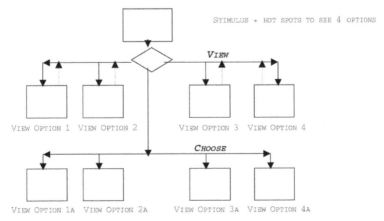

This is represented in Figure 3 (above), where the viewing options 1a to 4a, represent a presentation (which could be a graphic, audio or video file) which makes it obvious what the result of the learner's choice has been and whether it was an appropriate response or not.

Such a system can motivate the end user into a closer exploration of the various options and the reasons for making the choice(s). That this is a long way from a 'standard' MCQ is fairly obvious, but in terms of computer interactivity (as opposed to instructional interactivity) there is structurally very little difference. We are now into the realm of *conditional branching*, or the "What happens next if I make this decision?" context.

When *Expodisc Spanish* was created, (Bangs & Staddon 1989) forms of feedback were used which had to be learnt from other disciplines, due to the lack of previous experience in the language field on which one could draw. One of the most valuable of these forms in our view, (and this is a view which we still hold today), was conditional branching (Davies 2001). Summarising briefly, this technique implies the presentation of material with decision points, at which the user makes a choice, with or without the possibility of viewing partial or whole consequences of the choices to be made, and then is led down a path which makes it *intrinsically* obvious that the decision made was wholly or partly correct, or an error has been made.

Conditional branching, almost by its very nature, can offer a more intrinsic type of feedback, and, if coupled with a sympathetic production of content, can engage the learner and make them feel much more involved in the action, in itself once more a motivating factor. Conditional branching obviously has many uses other than in simple feedbacks to activities and exercises, which cannot all be considered here, but it is often neglected when designing multimedia for language learning. There are probably several reasons for this. The obvious objection is that it can impose a tight framework on the structure offered to the student, and might be seen, at a pedagogical level, to conflict with a constructivist approach. But given the realisation that the learner often needs 'scaffolding' even in task-based learning, this does not seem to be too much at odds, if carefully designed. The other main objection is simply that, whilst easy to program, it is difficult to design and costly to produce the increased amount of content needed.

Unfortunately, such techniques are rarely seen, perhaps because they can be costly on resource content acquisition, and relatively difficult to design in instructional terms. An exception is the recent development of the *Sunpower* course (Bangs et al 2000). But the lessons can be applied to simpler situations, as has been shown in the above MCQ examples.

Individualising feedback

Even so, the above techniques do not in themselves answer the need to have 'adaptive' feedback in language learning programs. To examine this aspect a little further, one can imagine the ideal scenario. A learner works on an exercise, and makes an error. The feedback he or she receives is totally 'tailored' to the individual student – such as "Sheila, you got the same thing wrong last week – revise what we did in class last Tuesday or email me if you don't understand it". Even better than this would be a more intrinsic approach where the presented feedback is not a meta-level message.

Obviously, the likelihood of achieving such a level of detail, unless engaged in one-to-one teaching, is quite remote. But the principle holds good. The ideal of matching the feedback to the individual learner is not, of course an easy proposition. This is in part because commercial programs aim at the widest possible audience for obvious reasons, and very few of them include an element whereby the feedback can be tailored to the specific target audience or individual learner profile. This is a pity, because in fact it would not be too difficult to include an element of the program which could, optionally, be made 'open' for the tutor to add or amend lines of feedback. So it will probably only be when authoring learning programs that this possibility can be included. Even then, the tendency is to allow for the author to use the system to create a defined routine which is then encrypted for delivery and has no re-usable components. Some work has been done on the creation of a system which makes a clear separation between the functionality of the program and its content resources, (not unusual in itself) but which also provides an intermediate text file component which allows for customizing as well as being the key to a tool for very rapid re-versioning or additional authoring. The text file represents a 'script' which contains references to all or any of the content resources, and either carries feedback lines themselves or refers to text or other multimedia files as feedback. Thus the author has no need to enter the program itself in order to adapt the content being addressed – all that is needed is to re-write or edit a text file, which may well be stored in a menu of resources which the end user can call up as needed. The program takes lines or strings from the text file as it needs to in order to run the program. Unfortunately such a system has yet to reach the market, but a sample file might look like this in part:

```
[Stimulus] "What is the correct phrase for saying Good morning
in Spanish?"
[Option1] "Buenas tardes"
[Option2] "Buenas noches"
[Option3] "Buenos días"
[Option4] "¡Hola!"
[FB1]  "No, this is used in the afternoon. We dealt with this
last week, so I suggest you revise your notes"
[FB2]  "No, this is used in the evening and usually when you
leave or go to bed. We dealt with this last week, so I suggest
you revise your notes"
[FB3]  "Yes, that's it – remember that morning lasts until
(Spanish) lunchtime."
[FB4]  "You can say that, but it really means Hi, and is
informal"
[FBAudio1] btardes1.mp3
[FBAudio2] bnoches1.mp3
[FBAudio3] bdias1.mp3
[FBAudio4] hola1.mp3
```

Should the teacher wish to change the lines of feedback, all that is required is to re-write parts of such a text file, including, if necessary, the name and location of the audio or other multimedia file. Such a system is also useful for the rapid authoring of large quantities of material within a project or in a collaborative venture.

So the only real way to even approach this level of detail has to be through the use of authoring tools. Many of these, even today, do not allow for more than generic, adapted feedback messages. If this is the case with 'traditional' delivery systems such as CD-ROM then how much more is it true of web-based systems? Most language tutors will not be willing to learn how to program in Flash or Java, any more than they would have been ready to learn BASIC in times gone by. Until all the appropriate authoring tools are made available, the important thing, in our view, is to hold fast to good pedagogical principles, and be able to create *instructional designs* based on the knowledge of what is or is not possible before passing the design work to programmers.

It is hoped that more of the authoring routines we all need will appear before too long. In their absence we are left with just the possibility of creating discrete programs and routines. Without going into too much detail about the need for authoring systems in general, it is probably well accepted that without some form of authoring program production costs become very high. All users of computers at some stage or other use an authoring program, often without realising it. If web pages are created without writing directly into code such as HTML or XML, then an authoring system will have been used. The classic argument offered for using authoring systems is that they reduce the ratio of development to student use time. To put this into context, the time taken to develop one hour of learner usage time may be frighteningly high if programming without such a system – in the order of hundreds to one. A good authoring system should considerably reduce this ratio.

Many different types of authoring systems are used to create learning programs. Some of the most powerful were not developed specifically for language learning, but have been successfully used for this field. Software packages such as Authorware, ToolBook and Director have been on the market for many years. The problem with them is that, although they do simplify programming and are extremely powerful, they still involve quite a high learning curve if one is to get beyond the more basic presentations, and in some cases are expensive. (Davies deals with the training issues in chapter ten). One answer has been to develop sub-sets of routines using such programs, as was the case with the 'scripting' example shown earlier, which used Authorware, and another system was developed using Director. Authoring systems more specific to language learning have also been on the market for many years. These tend to concentrate on specific sets of tasks – questions, MCQs, text manipulation, and others, and include Wida Software's well-known package *The Authoring Suite* and Camsoft's *Fun with Texts,* among others. (See Bangs, 2002).

Feedback, the Internet and templates

It can be said that the use of such a mixture of tools and the arrival of useful multimedia and extensive storage facilities such as CD-ROM began to see the development of a sound approach to pedagogical and instructional design, with good practice abounding in many centres, and excellent materials being produced. The past tense is used because then a new phenomenon came along which has turned the language learning materials production world on its head – the growth of usage of the Internet and its associated expectations. This requires some explanation.

The enormous 'hype' over e-commerce and the vertiginous growth of the so-called 'dot.com' companies has not left the world of learning unaffected. The demand for 'e-learning' has grown amazingly in recent years. The problem is that this term has never been satisfactorily defined – or rather has been defined in very many ways. This is not the forum to go further into these aspects, but suffice it to say that there is a demand in the world of education and training for learning to be delivered online over the Internet. The rush to provide web-based 'solutions' to learning situations has (with, of course, many honourable exceptions), led to a lack of application of good pedagogical and, above all, instructional design principles. The very nature of the web and the requirement for the use of HTML for the range of browsers, has often been cited as a limiting factor in the design of web-based programs. But even without the use of additional routines such as Java, Flash, JavaScript and so on, it is really quite easy to provide meaningful, and even intrinsic feedback with the use of the simplest of authoring tools such as Front Page. The pedagogical poverty of the endless series of MCQs with three or four options, one distracter, always only one correct answer, and the only feedback a score of x/y is such a common sight that it is difficult to understand why those creating such routines are satisfied with the results.

It is true that the Internet, and more specifically the World Wide Web suffers from a major deficiency which, paradoxically, is also its strength. The acceptance of the common standard for browsers has meant that one is faced with the challenge of delivering materials which (more or less) have to be read with HTML web browsers. The problem with HTML as a programming language is its basic simplicity as a hypertext system. To achieve the complexity of instructional design which we were able to produce with CD-ROM based materials is extremely difficult for web-based routines. One has to use other programs, such as Flash, Dreamweaver and so on, to add to the functionality of basic HTML routines, and the learning curve for such systems is often very high. There are very few authoring systems for learning routines which offer true web-based compatibility. Those which do exist use complex languages like Java and JavaScript (not the same thing) to enhance the capabilities. One of these is the familiar and successful *Hot Potatoes* (see Arneil and Holmes this volume). Once its relatively simple systems have been learnt, it is really quite easy to create materials rapidly. Not only that, but the system offers the possibility of including good feedback routines, and manages to get away, for instance, from the 'classic' MCQ systems. Thus it can be seen that *Hot Potatoes* can answer some of the requirements for good feedback systems which have been outlined earlier in this chapter. The routines which *Hot Potatoes* offers are expanding and becoming more complex with further development, which is to be applauded. What they offer in effect is a ***template***. This term is sometimes used synonymously with a ***shell***, into which content can be placed very easily and to good effect. Templates are a form of authoring system, and the most basic are rigid containers which offer little or no possibility of affecting the functionality of the routine. *Hot Potatoes* goes beyond this by allowing some input into the graphic and functional design of each of their templates, to good effect. If one merely considers the MCQ routine, then it is clear that it is an easy matter to commission the writing of significant quantities of content, delivered in a standard text format, and including appropriate feedback lines, more or less individualised for the context, which can then be fed very quickly into the template. Multimedia files are also possible (and often desirable).

At this juncture one could stop and ask why, given the possibilities that *Hot Potatoes* and others demonstrate, one sees so many examples of bad practice on web sites. It would be invidious to quote any, but it is not difficult to find examples, for instance, of very basic MCQ routines with the standard text options and no feedback other than scoring, some of them on what might be considered quite prestigious sites. The constraints mentioned earlier concerning web browsers and HTML do not even apply here, as it is perfectly possible to create basic routines with good feedback even using standard HTML programming or authoring systems, such as Front Page. Indeed, the web can cope with almost every type of feedback mentioned in this chapter, including conditional branching, even though without some of the sophistication and ease of use of *Hot Potatoes*. What seems to be lacking in fact is the implementation of good instructional design rather than any technical weakness in programming. This leads one to ask the question – why should the transfer of our learning platform to the Internet require us to accept a lower standard of quality in pedagogical and instructional design than that which had been hitherto developed and honed?

One theory which may explain this situation is that the introduction of web-based 'solutions' has brought about a rupture between pedagogical and instructional design which was much less prevalent when the concentration was on packages to be distributed by other media such as CD-ROM. Experience has shown that whereas in the latter case many practitioners were willing to use the authoring tools available to create volumes of material, the situation created by use of the Internet has been that it is not so easy to find authoring tools that do not require a steep learning curve which few teachers have the time or willingness to undergo. The implication, therefore, is that much of the creativity has been passed to technologists who have the necessary technical and programming skills to create the routines, but without the pedagogical background to create sound programs. But this still does not explain why the originators should be satisfied with the results. Perhaps they have been convinced by the technologists that this is all that is possible using web-based technologies. Or perhaps (more likely) it is all down to a question of costs. Whatever the real reason, it does seem as though the 'baby has been thrown out with the bathwater'!

The need for instructional design

The key to all this should be the role of the instructional design process. In our opinion it is this stage in the design of any multimedia program which is absolutely crucial, but which often seems to be overlooked (Orellana et al 2001). And this, we would argue, is not unique to web-based learning, though it is more transparent there. A good instructional design forms the bridge between the pedagogical approach and academic design, and the interactivity of the program itself. Without it any project is left open to the risk of either poverty of interactivity or inadequate pedagogy. Returning to our specific theme of feedback, it is at the instructional design level, whether designing for the web or CD-ROM, that the role of feedback can be developed and integrated. It could be argued that there has often been insufficient attention paid to this part of the design cycle over the years in the language-learning field, and that it is not just a recent phenomenon. In the early days of multimedia, when the big silver discs were almost the only way to present true video and graphic images, albeit with the need for costly interactive video machine configurations, there was virtually no previous experience in multimedia designing for language learning. Those engaged in the field had to look outside the world of language learning for information and inspiration (Bangs & Staddon 1989). Our view is that such an approach would benefit the quality of production even today.

MALTED – a partial answer

The experience of working with authoring systems, templates and scripting functionalities fed into the setting up of the *MALTED* project. The acronym stands for *Multimedia Authoring for Language Tutors and Educational Development*. With partial funding from the European Commission's *Educational Multimedia Task Force* program, the project was developed with a total budget of some €3 million. The major objective was to create a user-friendly authoring system, which, though based on the provision of a wide range of template structures, could nevertheless offer another level of linking routines together to form a coherent piece of courseware, somewhat along the lines of WebCT discussed in chapters three and four. Whilst *MALTED* does not offer all the functionalities which might be expected from a CD-ROM application, it is a wide-ranging authoring system which outputs into an XML format, and which uses Java to make available more complex functions such as drag and drop which are not usually available in 'simple' HTML. The consortium creating the system has carried out extensive trialling with language teachers, especially in Spain, where the Ministry of Education has been at the forefront of developments. Features have been revised and amendments and additions to the routines have been incorporated, and *MALTED2* has been launched as freeware, available for download.

The relationship between *MALTED* and the question of feedback is fundamental, and relates to the enormous flexibility even within the template design itself. Although some training in the system is needed (an online tutorial is available), a simple drag-and-drop interface allows for the inclusion of any media object on to any screen, and all media objects carry the possibility of interactivity of many sorts, including the provision of context-sensitive feedback. There is also a 'blank' template which can be used for any type of 'clickable' interactivity and is one of the most useful parts of the system, especially as the designer can place conditions on the destination of the feedback from any clicked item. Other templates include: various styles of MCQs, true/false, matching in various guises, several types of gap filling, crosswords, hangman, memory game, ordering, dialogue listening, recording and playback, menu routines and some differing open ended panels with email facilities. Further work will add to these facilities. The ability to re-size, cut, copy and paste objects at all levels assists rapid creation of series of routines, so that a well-constructed favourite 'frame' (the term for a routine created from the basic template), can be stored and re-versioned rather like a template in itself, but supplied with content or content markers.

A further powerful feature of *MALTED* is that it then goes one stage further, by creating not just a 'frames' level with discrete exercises, but a 'course' level, which links the frames into a hierarchical or linear piece of courseware, a powerful feature which adds an extra dimension to the system. The 'course' can link together frames created from the templates (or by adapting existing frames) with a variety of conditions (including time, score, and user variables) so as to create pedagogically complex and coherent learning routines without having to have recourse to an external Learning Management System. The effect which this should have on the question of feedback is an important one, since discrete exercises offer less possibilities of effective feedback, especially of the intrinsic variety, than longer sequences of linked activities.

There are developer and run-time systems (RTS), and the end user will have to have the RTS and Java installed. The really good news is that the whole system is free to use, with the condition that any further development of the template structure must be disseminated. Figure 4 shows a typical screen in editing mode.

The system is still being developed and enhanced. A major feature which will appear in forthcoming versions will be the inclusion of access to a remote asset base, from where resources and authored 'courses' can be downloaded for use and/or adaptation. Up-loading to the asset base will also be possible, subject to inclusion of the appropriate metadata (see Arneil and Holmes this volume, for more information about this aspect) and quality controls.

Whilst it would not be entirely correct to describe *MALTED* as a completely 'online' system in the generally accepted sense of the word, it nevertheless represents a major advance in the creation of web-based learning materials. As faster connections become standard, remote access to courseware created with *MALTED* will be more efficient.

Figure 4. MALTED editing screen

Conclusions and recommendations

The lack of pedagogically appropriate feedback is not a necessary concomitant of the change towards online learning. Many of the techniques which held good for routines produced for other delivery mechanisms such as CD-ROM, still apply when using web-based learning. Even with the simplest of tools, good feedback can be achieved with careful design. Whilst it can be recognised that time and financial considerations are important factors, these should not detract from the need to apply sound pedagogical principles to the materials being created. Part of the problem may be the broken link between pedagogical design and interactive programming, but even if the creators of a project or course need to hand over to the technologists for the actual creation of the routines, this should always be subject to a rigorous set of design documents which specify the exact nature of the interactions to be employed, including, of course, the types of feedback and the content to be included. Now that authoring tools such as *Hot Potatoes* and *MALTED* are becoming available, there is even less justification for failing to include powerful feedback systems, since these programs empower those on the pedagogical side to be in control of the design process. In this case the reference is not to 'design' in the graphical sense, as this may need to be added by professionals. Rather it is a question of what can be considered the 'missing link' which still needs to be addressed. This is the need for more awareness of, and training in, the instructional design stage of the multimedia creation process. If this is taken into account, and the new tools adopted, we should be able to see a new generation of pedagogically sound programs which nevertheless take full account of the technological advances at our disposal. Going one stage further, there is a clear need for the whole area of instructional design to be promoted – and in 'both directions'. On the one hand, pedagogical designers need to understand the interactions and other technical possibilities at their disposal, and which they can use to put appropriate learning theories into effective practice. On the other hand, the technology community needs to understand what the language learning community really requires from the application of the new technologies. The gap may be narrowing, but it still exists and ways should be sought to bridge it in the near future. In terms of concrete actions, the following recommendations can be offered:

- Consideration should be given as to whether web-based media will offer all the interactivity desired. If not, another format should be used – the computer and the Internet are not the only means of delivering learning, after all.
- The creation of a carefully formulated instructional design is essential. It may sometimes be necessary to look outside the world of language learning for additional inspiration.
- Feedback should be used that will offer meaningful learning opportunities for the student, and, wherever appropriate, should be of the intrinsic kind.
- Consideration should be given to the use of mixed media in order to enrich the feedback offered.

References

Bangs, P., Schroeter, F., Gove, P. et al (2000). *Sunpower, Communications Strategies for Business Purposes*, CD-Rom English Course. Munich: Hueber Verlag.

Bangs, P. & Staddon, S. (1989). Expodisc Spanish, Interactive Video Course for Exporters. London: NIVC.

Bangs, P. (2002). *Introduction to CALL authoring programs* (Module 2.5), http://www.ICT4LT.org/en/index.htm

Bationo, B. (1992). The effects of three feedback forms on learning through a computer-based tutorial. *CALICO,* 10 (1), 45-52.

Buscemi, C. (1996). *Feedback in Computer Assisted Instruction and Computer Assisted Language Learning.* http://www.edb.utexas.edu/mmresearch/Students96/Buscemi/researchproject.html

Davies, G. (1991). http://www.camsoftpartners.co.uk/expodisc.htm.

Felix, U. (1998). *Virtual Language Learning: finding the gems amongst the pebbles.* Melbourne: Language Australia.

Felix, U. (2001). *Beyond Babel: language learning online.* Melbourne: Language Australia.

Heift, T. (2001). Error-specific and individualised feedback in a Web-based language tutoring system: Do they read it? *ReCall,* 13 (1), 99-109.

Heift, T. (2002). Learner Control and Error Correction in ICALL: Browsers, Peekers and Adamants. *CALICO,* 9 (3), 295-313.

Laurillard, D. (1993). Program Design Principles, in *The TELL Consortium – Formative Evaluation Report*, available from http://www.hull.ac.uk/cti/tell/eval.htm

Orellana, N., Suárez, J.M. & Belloch, C. (2001). El diseño instruccional, una dimension clave insuficientemente atendida en la teleformación, in *Virtual Educa Online*, http://www.virtual-educa.net/

Pujolà, J-T. (2001). Did CALL feedback feed back? Researching learners' use of feedback. *ReCall,* 13 (1), 79-98.

Pusack, J.P. (1983). Answer-processing and error correction in foreign language CAI. *System,* 11 (1), 53-64.

Skinner, B.F. (1958). Teaching machines. *Science,* 128, 969-977.

Websites

Hot Potatoes – http://web.uvic.ca/hrd/halfbaked/
MALTED – http://MALTED.cnice.mecd.es

All websites in this chapter were verified on 18.09.2002

6

MOO as a language learning tool

Lesley Shield, The Open University, U.K.

Introduction

While many creative writing programs have made good use of MOO, recognising its potential to support collaborative and individual projects as well as reflective processes (English 1998, Kolko 1998, Harris 1996, Haynes & Holmevik 1996, Keenan 1996, Jordan-Henley & Maid 1995), this tool has been much less widely-employed in language education. Although there are several MOOs dedicated specifically to language learning and intercultural exchange (*Schmooze, Dreistadt, Le MOOlin Rouge, MOO Français, Mundo Hispano*), only a small number of language teachers and researchers has worked with language learners in the MOO environment. A growing number of research publications are appearing in this area but the apparent reluctance of language teachers to engage with MOO-based learning and teaching makes it difficult to identify clearly those aspects of MOO-use that exemplify best practice. For many learners and teachers, the very acronym *MOO* may appear frivolous and when expanded to its full title, *Multi User Dungeon – Object Oriented*, inappropriately redolent of the networked dungeons and dragons games of the 1970s in which it had its genesis. This is probably why MOO is now often known by the more respectable term *Multi User Domain – Object Oriented*, although it has retained some of the original, games-related terminology; participants are known as *players*, for example. As a result of its relationship to the dungeons and dragons role-playing found in MUDs (Multi-User Dungeons) and to the role-playing that is widely perceived to occur in social MOOs – and the occasional resultant bad publicity (see Dibbell 1993) – many educational institutions refuse to allow connection to any sort of MOO, educational or otherwise. Taken together, however, there are several reasons, both practical and conceptual, that make MOO a powerful and flexible tool for learning and teaching, able to support a variety of learning contexts and situations and to be adapted to suit different pedagogical approaches. These will be explored during the course of this chapter which is divided into two major sections. As well as defining the environment, describing how the tool can be employed and what the non-technical player really needs to know in order to use it successfully, the first section explains aspects of MOO that new users may initially find confusing or difficult. The second section introduces MOO-based language-learning research and draws upon several case studies to describe and evaluate different, MOO-based language learning activities, concluding with a reflection on the lessons that can be drawn from the research and how these can inform good practice.

Defining the environment: What is a MOO?

MOO exists in different forms, most notably social and educational, although there are also some role-playing MOOs (a comprehensive list of MOOs classified according to type, can be found in *Rachel's Super MOO List*). Since the early 1990s, MOO has been used by both educators for learning, teaching and training purposes and by communities of like-minded professionals to support research and development in their specific disciplines; *BioMOO*, for example, is 'a professional community of biology researchers connected to the Globewide Network Academy' (Rein 2002:5). In brief, a MOO is a public database housed on an Internet-enabled server. The database is accessed via telnet, a telnet MOO client, or, increasingly, a web-based MOO client. Essentially 'worlds of words' (Marvin 1995, truna aka j.turner 2001), MOOs are object-oriented programming environments and differ from plain text chat (henceforth 'chat') in that participants can use them not only to communicate synchronously with each other, they can also build their own landscape and objects within the MOO, using words alone in the construction:

> *MOOs offer every participant the opportunity to construct spaces and objects*
> *and to write code that in some way augments or increases the functionality*
> *with these virtual spaces. In this sense, MOOs are constructed social spaces*
> *in a dynamic process of continual evolution. (Thorne 1996:4)*

As participants' skills in using descriptive language and, in some cases, programming the MOO itself, develop, they can refine, redesign, enhance and expand their environment. Indeed, the architecture of any MOO is likely to be in a constant state of evolution as players who live and work there join and leave the community with objects they have made being removed from and added to the MOO accordingly.

What can be done in a MOO?

Writing in 1995 (p.8), Fanderclai makes clear her disappointment that imagination frequently appears not to take priority in MOO-based learning activities; '...all too often...' she finds educational MOOs using the spatial metaphor to reproduce the real-life campus and recreate the real-life classroom. Even now the virtual campus and classroom are predominant in educational MOOs, but MOO does challenge the traditional transmission model of learning. MUDS, and by implication MOOs, offer the teacher the means by which to control what learners do. MOO-adept teachers can lock learners into and out of certain parts of a MOO and control the order and type of synchronous communication in which they can engage. At the same time, these applications also provide the opportunity for learners to work in a self-directed way (Oren 1996, Fanderclai 1995). For many teachers, MOO-space can appear to be anarchic; players can engage in multiple conversations with different players in different parts of the MOO at the same time as apparently participating in 'classroom experiences' with their peers and their tutor. The initial shock of seeing text scrolling rapidly up the screen as players interact with each other and with their environment can be confusing and, occasionally, alarming. No matter how carefully a tutor plans a MOO-based class, the ludic nature of MOO, the sense of anonymity offered by donning an online persona and the way in which MOO releases the imagination can lead to chaos as learners take advantage of the flexibility offered by the environment. It is the potential to support learners in taking responsibility for their own learning and using their imaginations that makes MOO such an exciting tool for language learning and teaching. It is an environment in which players can adopt a persona different from that of the physical world, take risks in their use of language – and sometimes in their behaviour – and experiment without either losing face or sustaining physical harm. Indeed, MOO-related literature frequently refers to the way in which; '...students who are normally quiet and non-participators in a real world classroom often become much more active in a MOO class.' (Lee et al 1999:9) and there is some evidence that producing language in an environment that encourages inhibition may 'contribute to deeper learning rather than only toward greater time on task' (Hudson & Bruckman 2002:112). For differently-abled students, MOO can offer the same opportunities as to their able-bodied peers. Since MOO is primarily text-based, for example, screen readers can be used to read out the content of MOO-screens, and, because even synchronous MOO is subject to delay as players key in their responses, the visually challenged can operate on an equal footing to the sighted. Although the present author was aware of the equalising nature of MOO, this was underlined on one memorable occasion; in answer to a question about the reasons for his enthusiasm for MOO, an adult basic education student replied that he liked to use it because in the physical world he was confined to a wheelchair while 'in here, I can walk'.

What the non-technical player needs to know

MOOs can be accessed via the web, a special, standalone MOO client or raw telnet. Newer, web-based MOO interfaces adopt a point-and-click approach, while accessing MOO through a MOO client or telnet requires players to know certain commands in order to operate effectively within the MOO environment. Apart from the commands needed to connect to or disconnect from a MOO, however, it is only really necessary to know three commands in all to communicate with other players (Table 1).

Table 1. Commands required to connect to/disconnect from and communicate within
*   a MOO*

Command	Player keys in	Player sees	Other players see
Connect	Connect <playername> <password> Example: Connect Lesley xyzxyz	Player's connection message Example: Hello again, Lesley. Welcome back to the MOO.	<playername> + connection message Example: Lesley has connected.

Command	Player keys in	Player sees	Other players see
Disconnect	@quit	Player's disconnection message Example: You leave the MOO.	\<playername\> + disconnection message Example: Lesley has disconnected.
Say	Either: say How are you today? Or: "How are you today?"	You say, "How are you today?"	Lesley says, "How are you today?"
Page	page \<playername\> How are you today? Example: page Uschi How are you today	\<playername\> + page received message Example: Uschi has received your message.	\<playername's page message\> + message Example: You hear Lesley whispering in your head, "How are you today?" NB only the paged player sees the paged message
Emote	Either: emote smiles happily Or: :smiles happily	Lesley smiles happily.	Lesley smiles happily.

While it is true that the requirement to learn different commands places a heavier workload on novice players or *newbies* than the transparency offered by chat interfaces, it is noteworthy that such players may experience difficulty in using even the newer, supposedly simpler, point-and-click MOO interfaces. For example, the experience of Shield & Hassan (2002) working with a web-based MOO interface was that learners found the imposition of a two-dimensional graphical user interface (GUI) on a multi-dimensional virtual environment added to their incomprehension despite the point-and-click approach. Learners using the web-based interface, who could move around the MOO environment by clicking on hyperlinks rather than needing to key in the name of an exit from a virtual space (Figure 1, overleaf), reported that they did not understand the spatial metaphor of MOO, saying that they were unaware that not all players were necessarily in the same virtual space at the same time.

Although technical training in using the web-based MOO was provided, these learners were slow to understand basic communication modes, such as *say* and *page*, and the reasons for using one rather than the other, even though the GUI required them simply to click on the appropriate radio button to use the correct communication type. Once they had understood what was involved, however, mainly as a result of collaborating with each other and sharing their problems, they quickly became enthusiastic MOO-players, proficient in both MOO-based communication and object creation. MOO's reputation for having a steep learning curve before it can be used effectively may, then, rest upon more than the need to acquire knowledge of the appropriate commands; there certainly appear to be usability issues around web-based MOO interfaces which require further investigation.

Figure 1. Comparison of a standalone MOO client interface with a web-based MOO interface.

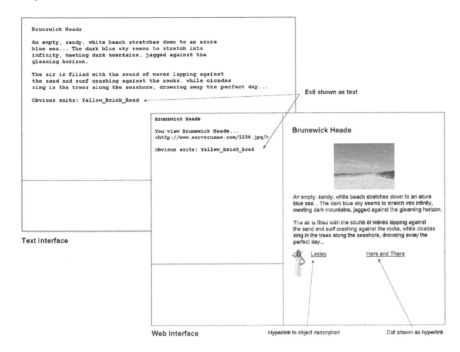

MOO: some common myths and misconceptions

MOO offers the opportunity to interact both in real-time and asynchronously with native and non-native speakers of the target language, but there has never been a great take-up of this technology in the language learning world. The reluctance of language teachers and learners to use MOO probably has its roots in some widespread though misplaced beliefs about the nature of the tool. The sort of questions that arise from this confusion will be examined in some detail before moving on to the ways in which MOO has been used by language learners and teachers and which demonstrate that MOO is a much more flexible and powerful tool for language-learning than many potential users may believe.

Is MOO just a game?

While it is true that some MOOs exist solely for the purposes of fantasy role-playing or for socialising, they are also environments in which relationships of various kinds develop between players; for example, Parks & Roberts (1998:527) report a study of MOO use in which:

> ...almost all MOO users (93.6%) ... had formed a personal relationship of
> some kind there. Participants had acquaintances, colleagues, friends, close

friends, romantic and sexual partners in the online world of MOOs. Close friendships were the most frequently reported relationships ..., followed by romances and friendships.

Each MOO is themed and has a purpose. Many MOOs have a website including details of their themes and purposes. This information is also usually included in the *mission statement* that appears on the opening screen when a player logs into a MOO (Figure 2).

Figure 2. The mission statement usually appears on the opening screen of a MOO

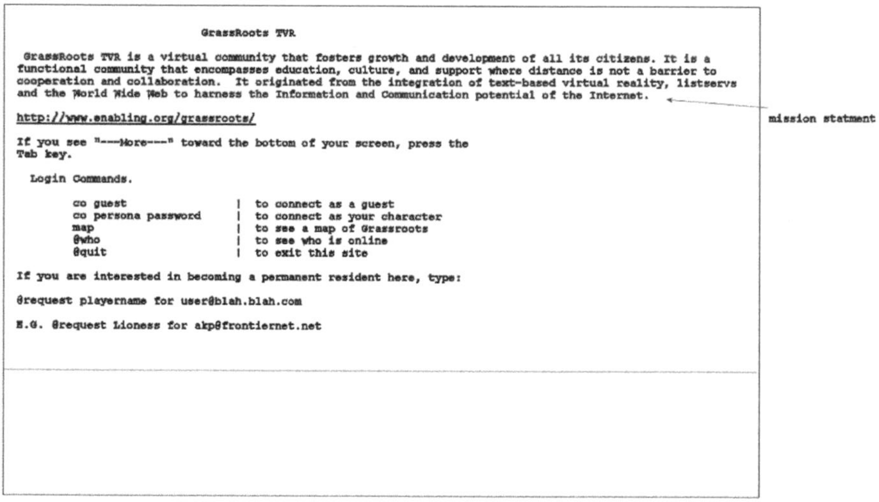

As well as this, the majority of MOOs, in particular educational MOOs, have other safeguards in the form of codes of conduct to which players must adhere. While these codes may vary, as codes of conduct in real world communities vary, they are strictly enforced, primarily by MOO administrators who may block access to the MOO by any player whose words or actions cause offence to other community members. In some cases, however, those who violate the code of conduct can be ostracised by other players and eventually choose either to conform or to leave the MOO. Although depending on the nature of an individual MOO and, in the case of an educational MOO, frequently upon the pedagogical theory on which it rests, the social structure of a MOO is usually well-defined and frequently hierarchical. There are different types of player whose functions within the MOO vary according to their class (Table 2). It is possible to progress through the MOO hierarchy, although players have to fulfil a series of stated, MOO-specific, criteria to do so. These criteria may be related to views of learning either as knowledge-transmission or as socio-cultural construct; administrators in one MOO, for example, may grant *builder* privileges as soon as a player writes a description for her/his MOO persona, while the MOO-administration of another MOO may refuse to do so if there are linguistic errors of any sort in the description, and may point out exactly what these errors are. In practice, unless a player has a specific project or goal in mind, it is not necessary to become a *programmer, teacher,* or *wizard.* The majority of players do not progress beyond the rank of *player* or *builder* and some are content to be *guests* for the duration of their use of MOO.

Table 2. The hierarchy of players in MOO

Player Rank	Description	Remarks
Archwizard	Has unlimited powers to create and destroy other players – including wizards – and objects	
Wizard	Unrestricted administrative powers. Can monitor what players are doing/saying at any time while remaining undetected	Depending on the MOO, there may be no difference between a wizard and archwizard
Teacher	Has the same privileges as a wizard but is unable to create generic players or monitor what others are doing or saying	This class does not exist in all MOOs; it is a type of enhanced programmer and can create characters for its own students instead of asking wizards to do so.
Programmer	Like a teacher but may be unable to create new characters. A programmer can create and program new objects and verbs (keyboard shortcuts to produce 'canned text').	The teacher/programmer distinction often does not exist.
Builder	The builder's privileges are limited to creating new objects and limited programming functions. S/he cannot, for example, make new verbs.	May be conflated with the 'player' character.
Player	Is unable to create new objects but logs in via a set user id and password.	Some MOOs allow the player the same privileges as a builder, or do not distinguish between the two classes.
Guest	Each MOO has a set number of guest ids. The first visit by any player to a MOO is as a 'guest'; this temporary persona allows the visitor a 'taste' of a MOO without committing her/himself to it.	Depending on the MOO, a guest character may be allowed to create objects while in a MOO; these objects will be recycled (destroyed) as soon as the guest character logs off.

MOO then, is not a game, although some MOOs may have a game-playing area where networked, text-based versions of games such as scrabble, tic-tac-toe etc. may be found; players do not usually score points for killing other players or opponents which may be *bots* programmed by other players. Often, there is a strong sense of community among players who may work together collaboratively to expand the MOO environment. As in a real community, friendships are forged and broken, relationships develop and disintegrate but there is likely to be a central core of players that is regularly to be found working and socialising on the MOO while other players join and leave.

MOO vs chat: the same or different?

On first inspection, real-time communication in a MOO (MOO-discourse) may appear to be very similar to real-time chat. Like chat, MOO supports synchronous text interaction which can be logged by participants, with the permission of all those whose exchanges are recorded in this way, so that they can read and reflect on these logs after the synchronous event.

On the whole, however, chat offers users an immediate and rather transparent means of communication; all that is required is to connect to an application such as IRC or an instant messaging service, to join a chat room, enter text into the text entry window and hit *Return* (Figure 3).

Figure 3. Typical MOO and chat interfaces

Participating in MOO-discourse is a little more complicated. It may be primarily for this reason that many learners and teachers who employ real-time, online interaction in their language learning and teaching choose chat rather than MOO. After all, why should they invest time in learning to use an application when the same thing can apparently be done much more simply and with less effort in another environment? The answer to this is simple – MOO provides features that chat does not (Table 3).

Table 3. Major differences between MOO and chat

Chat	MOO
Uses a 'rooms' metaphor. Users move between rooms by selecting from a (drop down) list.	Uses a 'spatial + rooms' metaphor. Players move between rooms by following exits in room descriptions or by *@joining* other players in different rooms. Some MOOs provide players with maps to help them navigate MOO-space.
Users meet in 'rooms' or 'channels'. These are featureless, animated only by the online presence of users.	MOO is *object oriented*. Players are able to create their own objects, including rooms, which they can describe and program. Objects can be viewed and interacted with by all players in a MOO, regardless of the creator's online status.
While some may be permanent and public, users can create their own, private rooms while online. These usually cease to exist when the creator is no longer online.	All MOO rooms are permanent. They may be public (part of the general MOO environment) and open to all players, or private (created by individual players and perhaps restricted to a specific group of users). MOO is extensible by its players. Unless a player ceases to frequent a MOO, any objects s/he creates are permanent until s/he recycles them her/himself. When a player leaves a MOO community, the administrators may delete that user's objects or, if they are considered to be more generally useful to the MOO community, they may be retained as part of the public MOO domain.
Supports: synchronous chat. **May support:** instant private messaging between users, file exchange (audio, graphics, text) discussion lists, message boards, games (often card games), sharing urls, voice chat, emoticons (pre-made graphics used to express emotions), font styling, logging functions, text-to-speech. **NB:** Instant messaging, file exchange, discussion lists and message boards each occur in a separate window. Users may have to manage many open windows at once.	**Supports:** synchronous chat, private messaging between individual players, emoting (expressing emotions in words), logging functions (usually built into a MOO client rather than into the MOO itself, although *tape recorder* objects can be made by players to record their conversations), MOOmail (a type of MOO-based email system), discussion lists, message boards, games (usually word games or text versions of card games). **May support** (depending on the MOO or MOO client): sharing urls either via imported web pages or, for the visually challenged, via the importation of text which can then be read out by screen reading software, collaborative description and programming of objects, 'file exchange' (objects can be programmed to allow players to view (streamed) audio, video, graphics and text files).

A text-only technology?

MOO is often referred to as 'text-based virtual reality', so it is not surprising that it is usually thought of as having a text-only interface. This is, however, no longer an entirely accurate description. Although some MOO clients still support a text-only interface, some now also support hyperlinks to multimedia files. Web-based MOOs may allow players to incorporate sounds, graphics and text into their MOO objects and may enable collaborative web-tourism through the sharing of urls. Some web-based

MOOS may even support players in designing and developing shared objects and writings together. There are, however, advantages to a text-only interface:

...the sensorial parsimony of plain text tends to entice users into engaging their imaginations to fill in missing details while, comparatively speaking, the richness of stimuli in fancy virtual realities has an opposite tendency, pushing users' imaginations into a more passive role. (Curtis 1992:17)

It may also be true to say that the language produced in MOOs that are not solely text-based differs from that which occurs where text is the only medium of expression (Figure 1). This is, however, currently a little-researched area, and worthy of further investigation beyond the scope of this chapter.

It should further be noted that, unlike chat, MOO supports asynchronous communication, including tools such as MOOmail – a type of in-MOO email system – and objects called 'notes' which can be programmed to be journals, books, and so on, according to a player's wishes (for those players who are unable to program in MOO, generic 'journals', 'books' and so on are often provided so that they can take advantage of the possibilities offered by such artefacts).

Written, spoken or a hybrid? What is the language of MOO?

Communication in a MOO, both synchronous and asynchronous, takes place via text. This may lead learners and teachers to assume that only written language can be practised and that MOO is therefore best-suited to developing (creative) writing skills. In fact, the exact nature of synchronous online text exchanges of all sorts has been debated for many years. The consensus amongst researchers is that such communication is neither spoken nor written in nature but, rather, a hybrid of spoken and written language (Crystal 2001). Levy (2000:185) implies that the exact nature of the language produced is perhaps not at issue. What is important, he suggests, is the learner's purpose in studying the target language; if this is to communicate in face-to-face situations, then environments such as MOO that allow synchronous communication are '...better regarded as rehearsal for real interactions...'. For those learners whose communications with native speakers of the target language will always be online, competence in online communication will '...become their declared aim...'. While the debate continues about the exact nature of synchronous chat and whether transfer from one medium to another takes place, as suggested by the work of Chun (1998) and Schwienhorst (1998b), what does seem to be emerging is the fact that, at worst, working in MOO provides learners with the opportunity to '...practise their skills in authentic two-way communication...' (Kötter 2001:290) in a way which reflects the language produced by L1 speakers in the same environment (Weininger & Shield 2001, Shield & Weininger 2002a, 2002b, 2002c).

MOO pedagogy: what can we learn from research?

While MOO has not been widely used for language learning and teaching, teachers have been making suggestions about possible uses for some time (Turbee 1995, 1996, 1997) and there is an increasing body of research into using MOO to support the learning and teaching of both modern and ancient languages (Schwienhorst 2002). Such research may be roughly categorised according to learning contexts and learning activities (Table 4). There is, of course, a degree of overlap in terms of learning contexts and situations; a series of individual learning activities such as solving a treasure hunt or participating in an online discussion group may lead to an overall learning outcome which falls into a broader, over-arching category.

What appears to underlie all these studies is a pedagogy that stresses a constructivist approach to learning (see Papert 1993, Brooks & Brooks 1993) rather than the teacher transmitting knowledge to the learner. In such an approach, the learner takes an active part in the learning process, building up or constructing knowledge and understanding step by step either individually or collaboratively with peers. Furthermore, learning is seen as a social construct where learners interact with other learners and/or native speakers of the target language, adopting Vygotsky's notion (1978) of the importance of collaboration and social interaction to the learning process (see Felix this volume).

Table 4. Learning contexts, situations and activities frequently represented in MOO-based research.

MOO-based learning context/situation	*Learning activities*
Tandem: two groups of speakers of different languages use MOO to work together to achieve defined outcomes. (Schwienhorst 1997, 1998a, 1998b, 1998c, 2001, Donaldson & Kötter 1999a, 1999b, Kötter 2001). Learners on two sites work in groups of two, three or four. Each group contains at least one native speaker of each language and at least one member from each site. During MOO-based tandem projects, participants are expected to adhere to the three principles of tandem learning, *reciprocity, bilingualism* and *autonomy.*	Typically, activities require the learners from each site to reach a specified outcome, which may be, for example, a jointly produced piece of writing, a synchronous, MOO-based presentation on a specific topic, a part of MOO-space representing a real or imagined town, and so on. Learners may be involved in interviewing each other synchronously and asynchronously within MOO, jointly devising questionnaires, searching for and sharing information or collaboratively creating, describing and programming objects and MOO-space.
Project-based: groups (from around the world) comprising native and non-native speakers of a single language (Davies et al 1998, Shield & Weininger 1999, Shield et al 1999, Shield et al 2000, Shield et al 2001a, 2001b) Learners who are geographically distant from each other work in groups. If there is an imbalance in the number of learners from different locations, a group may include more than one, but no more than two, learners from the same site. Teachers offer more or less guidance, depending on learner language level, about the steps required to achieve the identified goal. The notion of *audience* is important and project outcomes are generally available beyond the immediate peer group of learners and tutors involved.	Within a framework provided by the teacher/ instructional designer, learners negotiate with each other and their teachers in the target language to define a specific learning outcome they wish to achieve and how they intend to achieve this. This outcome may involve learners in activities such as researching a particular topic and developing a web-page collaboratively using MOO as the research and discussion space, finding out more about a historical period or geographical area and creating a section of MOO-space to illustrate what has been discovered by the group or building an in-MOO puzzle for others to solve.

MOO-based learning context/situation	*Learning activities*
Simulation: learners role-play a specific situation within a simulation framework provided by the teacher/instructional designer. They build their environment and act out their MOO-lives in character using the target language (truna aka j.turner 1998, Shield & Hassan 2002). Learners may all be physically present in a computer lab or at a distance from each other and meet only virtually in-MOO. Because the MOO-space for the simulation belongs to the learners, the role-play may deviate from the scenario provided in the original framework, as the MOO-personae take on their own lives and participate in events that were not predicted by the teacher/instructional designer.	Employing the target language, learners work individually to create their online personae and co-operatively and collaboratively to build their own part of MOO-space, using either their imaginations or researching real places and adapting these to their MOO-space. Once the area has been built, learners take part in a role-play which may comprise a single simulation activity such as a meeting or an extended role-play.

Tandem MOO

Schwienhorst (1998a) provides a concise overview of the history of and concepts underlying tandem exchanges. Initially, he points out, these were face-to-face encounters between two speakers who were learning each others' mother tongues. By 1994, these tandem exchanges were implemented via email between learners from different educational institutions. The concept of tandem exchange is based on three principles; first, *reciprocity*, where each learner receives as much from the partnership as s/he donates to it; secondly, *bilingualism*, where the mother tongue and the target language are used equally by all participants; thirdly, *autonomy*, where each learner is responsible for her/his own learning.

While tandem email exchanges have been very successful, MOO has been much less used as a vehicle for tandem language learning projects, its main advocates being Schwienhorst (1997, 1998a, 1998b, 1998c) and Kötter (Donaldson & Kötter 1999a,1999b, Kötter 2001), although research has shown the MOO environment to be a powerful tool to support such partnerships. For example, reporting on a tandem exchange between learners in Germany and the US, Kötter (2001:301) remarks upon the beneficial effect of synchronous communication between tandem partners; they can, he points out, receive immediate feedback and help from each other, rather than waiting for an email response. Learners in that study, he notes, did adhere to the second principle of tandem exchanges and used their mother tongues and the target languages in approximately equal proportions, although, perhaps tellingly, they tended not to correct each other as much as their tutors had expected.

> *... it seems that the continuation of the discussion of the topic in hand was often more important to learners than the provision of feedback on errors which hardly ever affected their understanding...*

Collaborative project work

While tandem MOO projects have involved learners in collaboration on specific learning activities and outcomes, their focus has tended to be on the tandem rather than the collaborative aspects of the undertaking. The outcomes of learners' work in tandem MOO exchanges are not generally available to a wider public. Shneiderman (1997) proposes that audience is an important motivational factor in learning and that learning outcomes should be achieved by participation in meaningful activities. Drawing upon this approach, some MOO-based language learning projects have made use of the possibilities the environment offers for collaborative working, focusing on the opportunities MOO offers for such collaboration and stressing the notion of audience when supporting learners as they work towards their identified goals. One such project is reported by Shield & Weininger (1999); learners and teachers from Japan, Brazil and the US were recruited to participate in a project whose major outcome would be the production of a collaboratively-written group web-page with English as the common language and whose topic would be related to a special area of the MOO in which they worked. This web-page would be publicly viewable, providing an audience beyond that of immediate peers and tutors.

The initial attempt to launch this collaboration was unsuccessful; participants were unable to meet because of technical problems or time zone differences (this led to little socialisation between either learners or teachers which, in turn meant that communities of learners did not form, and thus to demotivation resulting in a high attrition rate), there was no single member in each group of learners who had sufficient experience of the MOO environment to support her/his peers and some of the participating teachers did not subscribe to the constructivist philosophy underlying the research. Reflecting on what had been learned from the failure of the first iteration of the research project, Shield & Weininger ensured the second iteration involved students located in similar time zones to each other, Australia and Japan, and that each group of learners included at least one member with advanced MOO skills. This time, the outcomes were more successful; groups, chosen by learners, formed and socialised; learners communicated with each other both synchronously in the MOO and asynchronously via MOOmail. Not all groups, however, developed the web page which was the intended outcome of their collaboration, although logs from the group discussions revealed that all groups talked about the possible content of their web pages. Interestingly, although the focus was not the same as for Kötter's tandem exchanges, learner behaviour does appear to be similar – again, topic discussion, rather than concrete, public outputs, seemed to be the most important factor for learners.

Building upon the outcomes of Shield & Weininger's work, another collaborative MOO project was set up to explore further the motivational effects of the notion of audience upon language learners. Participants from Australia, Korea, Ukraine, Japan and Sweden were involved. This time, the focus was upon the creative aspects of the MOO environment – learners initially worked alone to create their own, individual 'rooms' in the MOO. Once these rooms were complete, learners chose where to attach them in a specially-built area in the MOO. In their groups, comprising at least one member from each participating institution and at least one member who was experienced in the use of MOO and thus able to support her/his peers, they evaluated each others' rooms. The intention was that they would suggest improvements, both to the language used in room descriptions and to the creative concepts themselves. Some learners, apart from those from the Korean institution who were able to visit the MOO less often than other participants, created individual rooms and took pride in showing them to others (Figure 4).

Figure 4. Learners discuss a room

```
Sarah says <To Jo> "Well like my room"
Jo says, "Sarah " This is a very nice room!! ""
Susie says, "yeah, this is a very nice room:)"
Sarah <to Louise> "it descirbes a train station"
Jack says, "Sarah It's a nice room. I'm very happy being
here."
```

It was noticeable that although learners liked to show their rooms to each other, they did not exactly enthuse about the rooms they visited. However, when they moved away from discussing the learning activity to a 'real world' or authentic topic they often became more animated, volunteering information and asking questions of each other about cultural aspects of their lives, for example (Figure 5).

Figure 5. Learners discuss their eating habits

```
Sarah says, "I got to have my lunch."
Jim says <To Sarah> "Oh, what are you going to eat?"
Sarah says <To B> "a sandwhich."
Mark says, "Sarah "what is your lunch??"
Sarah says <To Mark> "a sandwhich."
Mark says <To Sarah> "I am starving!! I envy you!!"
Mark says <To Sarah> "In Japan, it is ten o'clock in the
morning. So it is early to have lunch!!"
Jim says <to Mark> "What do you usually eat for lunch?"
Mark says <to Jim> "I usually eat rice, miso soup and salad."
```

Logs from this project revealed that learners tended not to evaluate critically the objects they and their peers were creating, despite the attempts of their tutors, both synchronously and asynchronously to encourage them to do so; logs from synchronous sessions show evidence of one tutor attempting to initiate such evaluation on several occasions, for example asking; "Would anyone like to ask any questions about X's room?", "I think it's a wonderful room, what do you think?" and so on. Another tutor visited every room created for the project and, by way of providing a model, sent evaluative comments of the type learners were intended to write to the owner of each room. The synchronous encouragement was partially successful, with learners making rather half-hearted attempts to talk about their peers' efforts at room-building, but the asynchronous model was completely ignored.

Like the preceding project, then, this collaborative approach to MOO only partially fulfilled its aims; many learners did create a room, some of which were very imaginative. These rooms can be viewed by any visitor to *GrassRoots MOO*. The lack of evaluative feedback by group members to each other is perhaps explained by two factors; firstly, logs reveal that learners formed friendships within their groups, socialising frequently between scheduled meetings. They may therefore have been reluctant to criticise their friends' work, seeing this as a destructive rather than supportive action. Secondly, Burdeau (1997) places MOO users into three categories, *Chat*, *Professional* and *Non-native speaker*, identified by linguistic and demographic features; the participants in the project described here fell into Burdeau's first and third categories being young and, mostly, non-native speakers of English. Examining the types of MOO-based activities engaged in by members of each of Burdeau's categories, Shield et al (2001a) found that reflective and evaluative discussions tended to form part of the activities of those who fell into Burdeau's second *Professional* category, that is older players, usually working in a professional area, rather than into either of the other two. If this is the case, then it is perhaps not surprising that learners in this project preferred to take part in chat-like activities rather than participating in reflective and evaluative consideration of their own and others' use of language.

While the notion of audience does not appear to have motivated these learners in the way that had been predicted, when compared to Shneiderman's learners, the differentiating factor appears to be one of distance; Shneiderman's learners were all physically in the same place, while the learners for both the collaborative MOO projects described here were at a distance from their MOO group partners and were possibly unable to meet as regularly as those in a face-to-face situation. It is also possible, of course, that the topic was of little interest to learners. While this is an area that requires further investigation, what is not at issue is that learners who participated in the MOO projects enjoyed meeting with their distant group members and that having a common topic gave them a purpose to meet in the first place; logs of their in-MOO discussions demonstrate that they participated enthusiastically in discussions/chat with each other, whether or not teachers were present (Shield & Weininger 2002c) even though these exchanges were more likely to be social than project-related.

SiMOOlation

Simulations can make use of MOO's tendency to encourage role-playing; learners do appear to take advantage of the relatively anonymous environment of MOO, sometimes adopting a persona completely different from their real life character, to take risks and to play with language, and this can be exploited for language learning purposes. A great advantage of MOO is that learners never need to 'step out of character' in a MOO-based simulation in the way that occurs in face-to-face simulations; the simulation can be a one-off event or it can continue over a period of time, with learners taking on the fictional role they have created each time they enter the MOO.

truna aka j.turner (1998) describes working with ESL learners to create a soap opera in the form of a virtual village and its inhabitants in MOO-space. Learners built the village and then took part in the life of that village in the characters they had created, reflecting on their adventures by writing a narrative after each MOO encounter. These reflections were then fed back into the MOO by posting them on virtual noticeboards in MOO-space.

Over a period of 12 weeks, Shield & Hassan (2002) ran a MOO-based *simulation globale*, that is an extended simulation comprising a series of stages and activities, with learners and teachers of French as a Foreign Language. The aim was to consider the viability of carrying out a simulation in MOO for older, distance learners who would never meet each other or their tutors face-to-face. The simulation scenario was that a British University had put out to tender a summer school to be based at a French University town. Learners worked in teams, with each team representing an imaginary French town, to build and populate their town and to put together a competitive tender to stage the summer school.

Learners for this project were slow to grasp the purpose of the simulation and some found it difficult to take on the persona of a fictitious character while still wrestling with the spatial metaphor of MOO. From an analysis of the evaluation questionnaires submitted by these learners, their difficulties seem to have surfaced mostly as a result of separating training in the use of MOO from the simulation; as a result, learners had to describe their own online persona for the training and then assume a new persona for the simulation. One learner in particular objected strongly to changing 'the name my mother and father gave me at birth' for a name suited to the simulation scenario and all demanded to know the rationale for this change of identity. Eventually, all learners accepted the need to take on a fictional persona, and one group even arranged to continue the *simulation globale* beyond its official cut-off date. Once learners had 'suspended disbelief' and begun to work in their fictional roles within the simulation, they soon became enthusiastic, spending many hours exchanging MOOmail – a MOO-based email system – and working together both synchronously and asynchronously to create their towns, always using the target language for their encounters. Indeed, the output of these learners far outstripped the researchers' expectations; by the end of the project, some learners had created a town newspaper and were jointly writing a news story about the ill health of the mayor of the town and its effects upon their lives. What was most noticeable was that learners had specific expectations of the scheduled sessions at which they met with their tutors, and that some of them did not think that these expectations were met. Feeding back to the researchers, learners reported that they had wanted tutors to take what they considered to be a traditional role and to correct their linguistic errors; they did not believe that tutors had done this. Rather, they said, tutors had been co-learners and had taken on roles within the *simulation globale* since overt error correction of the type learners expected would have required all the project participants to step outside its boundary and recognise their real as well as their assumed personae. Opinion about the effect of participation in the simulation on their communication skills in the target language was divided; some learners believed that communicating in real time in the simulation had stretched their knowledge of and improved their confidence in and ability to use the target language, whilst others thought they had made no linguistic gains. In fact, an examination of the logs for the simulation revealed that all learners had made some progress in their use of the target language. This progress can be classified in terms of the number of turns learners took in real time exchanges, the length and complexity of their input to both synchronous and asynchronous exchanges or a combination of both frequency and complexity of input.

While learners initially resisted taking part in the simulation scenario, the MOO provided the opportunity for them to work with each other and to form online friendships. The logs show evidence of group formation in terms of the frequency with which the pronoun *nous* (we, us) occurs both synchronously and asynchronously; the most frequent use of *nous* is found in the logs and MOOmail of the group that was the most successful in terms of output, even though this was the group that experienced the greatest difficulties in accepting both the MOO environment and the simulation scenario at the start of the project. These friendships and the development of a group ethos, in turn, appear to have encouraged learners to accept the challenge of the *simulation globale*. At first, their intention was to demonstrate to the tutors and the designers that they could (co-)operate successfully in the target language in the MOO environment regardless of how misguided they considered the learning scenario to be. Once they became drawn into the simulation, however, they became increasingly involved. They not only built their virtual town, but eventually became so absorbed in their virtual personae that they addressed each other only by their fictitious names, even when discussing events in the physical world, and adopted behaviour appropriate to their online characters. As a result, one participant – who was known to be polite and reserved in the physical world – became outspoken and somewhat aggressive in the virtual world, organising her team mates and complaining vociferously if they did not carry out their tasks to her satisfaction. On one occasion, she even accused the tutor of stealing the team's work and giving it to their rivals. While the tutor had, indeed, given some objects to the other team, this was an honest mistake and the learner's attack left her feeling that her personal integrity was being called into question in a way that impinged on her physical world persona. This incident could have caused the entire *simulation globale* to collapse, since relations between tutor and learners became strained to the point of near breakdown. The designers suggested to the tutor that she treat the attack as being aimed at her virtual persona and to incorporate it into the simulation scenario, congratulating her accuser on uncovering her dual role as an industrial spy working on behalf of the other team. This approach defused what had become a tense situation, as the tutor admitted publicly to being a double agent by the name of *Lien, Jaques Lien* (Bond, James Bond), returning the 'stolen' work with congratulations to the learner who unmasked her. As the *simulation globale* developed and learners became increasingly engaged in the lives of their virtual towns, they took the scenario in directions that had not been anticipated by either the designers or the tutors. Not only did they create a town newspaper but also, for example, a cyber-café which was linked to the website of a real French singer, an efficient town transport system and maps of their towns to aid visitors in navigating their way around them. They also made and hid objects in the MOO for their colleagues to find and supported each other when they experienced technical problems. The MOO environment in this case offered learners a space in which they were able to take control. Although their unexpected behaviour could have resulted in chaos, the careful scaffolding provided by the *simulation globale* contained the anarchy but allowed sufficient flexibility for learners to act autonomously and to use their imaginations to create objects and events of their own, thus engaging them in the use of the target language for an authentic purpose.

Reflections

Reflecting on the experiences of researchers working in the area of the use of MOO for language learning purposes, there are lessons to be learned which can inform good practice in the field. Both technical and pedagogical aspects of MOO must be considered when developing MOO-based learning activities.

The place of technical training

As Kötter (2001) notes, research has stressed the importance of providing learners with technical training in the use of MOO. Depending on the type of learning activity intended, the question arises as to whether any technical training should be rolled into the learning activity itself. How the technical training takes place may depend on whether the learners are physically in the same place or at a distance from each other and from their tutor. In the simulation carried out by Shield & Hassan (2002), for example, the learners were physically distant from each other, their tutors and the researchers and all MOO training had to take place online, making it very difficult to solve technical problems, although learners became adept at describing difficulties very clearly. Shield & Weininger (1999) found that for group work of this type, having at least one MOO-adept group member often addressed any technical difficulties, since learners were keen to help each other master the medium. Indeed, Shield & Hassan (2002) discovered that learners were much more likely to succeed in their MOO endeavours when they solved each others' technical difficulties than when they relied on their tutors or the project designers to help them. Rather gallingly, they all fed back that online help from the partner of one of their number had solved all their problems, this regardless of the fact that the help provided had been a reiteration of the training that had already been given by the tutors and project designers and which could found on the project-related web page!

The effect of practical and organisational issues

While the MOO environment seems to support a constructivist approach to learning, research suggests that successful MOO-based activities require careful scaffolding and advance planning. The effect of differing time zones can be problematic, for instance, especially where a project depends on group participation to advance its aims. Scheduling meetings between learners and tutors at appropriate times is therefore very important. Where group members are unable to meet synchronously, motivation for MOO-based learning activities is likely to wane; for group work, the optimal size appears to be around 6 members, since at least two members of a group are then likely to log in at an appointed time, while for tandem work, 'double dating' (1 or 2 learners from one site are paired with 1 or 2 learners from the other) addresses the issue of 'no shows'. It should be noted that the flexibility of real-time meeting scheduling is related to the type of learning activity involved; for example, the competitive element in Shield & Hassan's (2002) MOO-based *simulation globale* required that learners met with tutors in their 'teams' rather than logging in to any available scheduled tutorial regardless of participants. Since these learners were all located in the same time zone, this rather more rigid framework was relatively unproblematic, but for projects where different time zones come into play, it would be necessary to manage group membership much more carefully if learners were required to attend specific online meetings.

A common understanding of the underlying pedagogy and aims of a MOO-based project

Ensuring that all participants in a MOO project, both learners and tutors, understand and accept the underlying pedagogical theory and aims is vital to that project's success. Research suggests that where tutors do not agree with the aims of an endeavour or the way in which these aims are to be achieved, learners' experience will be unsatisfactory and attrition rates will be high, regardless of whether distance or campus-based learners are involved. Shield & Weininger (1999) report that the effect of a single tutor not subscribing to a constructivist approach to learning led to an entire group of learners withdrawing from their research project, while the feedback from learners involved in Shield & Hassan's work (2002) emphasised the importance of explaining aims and learning outcomes of the learning activity in real-time as well as via email and supporting web-based materials. In the latter case, one learner withdrew from the *simulation globale* as a result of not understanding its purpose or intended outcomes, even though these were clearly stated in all project-related materials, and several other learners reported that they would have found the concept of the simulation easier to understand had their tutors explained it to them during the course of a real-time meeting.

The roles of learner and teacher

> *A major dilemma ... when using text-based environments, is that the nature of the environment itself encourages a sense of freedom. This aspect has been noted by teachers, time and time again, as they watch their students leap into enthusiastic conversation and creativity, the open-ended theatrical script, and leave the restrictions of the classroom behind. (truna aka j.turner 2001:174)*

Although it is in the hands of the teacher/instructional designer whether a constructivist approach to in-MOO pedagogy is adopted (Hall 1998), most MOO research has assumed that although MOO-adept teachers can control MOO-based learner activity at least to some extent, it is an environment that changes and challenges the transmission model of learning so often found in online environments; learners can become increasingly autonomous in their choices of and decisions about learning. Of her first use of MOO with a class, Lasarenko (1996:1) describes a situation in which she spent time and care developing a virtual 'classroom' which faithfully mimicked the real classroom. Although it appears that she did not use the 'tools of control' referred to by Fanderclai (1995), she reports the experience so well-known to teachers who attempt to take a 'talking heads' approach to using MOO:

> *... after that MOO session my head was spinning. They didn't even look at all the handouts I gave them. And my lectures! If they couldn't even read the discussion questions in an adult fashion, what fate awaited those long,*

diligently prepared lessons? ... MOOs just didn't lend themselves well at all to a lecture-based form of teaching.

Research indicates that it is rare for MOO-based projects to meet exactly the learning outcomes defined initially by the designers of MOO-based activities; learners, either consciously or unconsciously, redefine those outcomes according to their own needs or perceptions, as evidenced in Kötter (2001), Shield & Weininger (1999) and Shield et al (2001a). Shield & Hassan (2002) found that learners took the initial *simulation globale* scenario and developed upon it, using their imaginations to breathe life into the town and characters they had created. Based on experience of simulations in other online environments (Hewer & Shield 2001), tutors took on roles within the *simulation globale* and, in character, made suggestions about how it could be developed. Their role was seen by learners as being one of co-learner of the uses that could be made of the environment and guide. Although they did, covertly, correct errors by reflecting language items back at learners in their accurate form, one participant complained:

> *In practice they helped very little indeed... and there were numerous quite basic mistakes being made constantly in all aspects of the work by students which were not corrected: the most common of these were probably incorrect agreements, mistaken genders of nouns, and incorrect verbs. While some such errors were typographical, others clearly were not, but there was no offer made to correct them and in the end I found myself offering to help correct... I felt quite uncomfortable at suggesting this, too, as the implication was that I was claiming linguistic superiority over my fellow-students.*

While learner expectations about the role of a teacher do not appear to have been met, emphasising the importance of making transparently clear the aims and goals of any MOO-based learning and teaching, the aim of the teachers and designers – to encourage learners to take more responsibility for their own learning – certainly were. This project also provides a very good example of the 'anarchitexture' underlying MOO and the way in which learners can control the direction a MOO-based project takes; as expected, learners repeatedly revised descriptions of their personae and their project area. What was unexpected, however, was that they also deleted objects and MOOmail without notice, making unilateral decisions about what was and was not pertinent to the project, thus displaying a high degree of autonomy as learners.

Conclusion

MOO is a very powerful tool whose potential for language learning is not yet fully understood. It is probably too soon to identify a single 'best-practice' approach to designing MOO-based language learning activities, but, used appropriately, MOO can facilitate access to native speakers of the target language and address differences in students' backgrounds, interests, needs, learning strategies and abilities (see Felix this volume). Much work remains to be done, but reports from teachers and researchers strongly suggest that the MOO environment lends itself to a constructivist rather than to a transmission model of learning, with learners working collaboratively with each other and their teachers to achieve agreed outcomes. Because MOO allows learners to describe and develop their own MOO-personae, they are provided with a degree of anonymity that allows them to take risks in their use of the target language without losing face. This may result in learners' inhibitions in using the target language being lowered, resulting in 'greater linguistic output' in the target language (Hudson & Bruckman 2002:112). Being network-based, MOO affords learners the opportunity to interact with native speakers of the target language in a way impossible in a face-to-face classroom. Furthermore, activities that could ordinarily be considered banal become meaningful when learners communicate with each other across time zones and geographical boundaries. For example, while a discussion of the weather with learners in the same physical space may be of little interest, the same activity taking place between learners based in, say, Ukraine and Japan, takes on new meaning; it is engaging because the difference in interlocutors' location provides an information gap they are likely to wish to bridge. MOO also supports cross-cultural exchange; learners and native speakers can share information and can even develop virtual representations of agreed aspects of the target culture together, thus becoming 'part of dynamic cross-cultural knowledge building communities' (Peterson 2001:447). As well as this, MOO offers tools that provide for the construction (by learners) of pre-programmed CALL-type materials for other players, such as the 'grammar maze' that can be found at *SchMOOze University MOO*. In this way, learners are able to construct their own knowledge about specific aspects of the target language at the same time as sharing this knowledge more widely by collaboratively designing and building artefacts with the purpose of donating them to the MOO community as a whole. The different types of synchronous and asynchronous interaction supported by the MOO environment may even contribute to the development of both the autonomous language user who employs the target language to communicate with native and non-native speakers (where the target language is the only shared language available) and to the development of the autonomous language learner whose writings are permanently available for all to see (Schwienhorst 1998b) and who is motivated by this fact to reflect on and revise her/his target language output with the help of peers, tutors and native speakers.

MOO, then, is a promising yet complex environment. It can be difficult to harness this environment to achieve real learning outcomes but this can be overcome. What is required is commitment from both learners and teachers to familiarising themselves with the essential technical aspects necessary to function in the MOO environment and acceptance of a pedagogical approach that emphasises the social and collaborative aspects of learning where the teacher's role is one of co-learner and facilitator. For the language professional interested in exploring possible uses of MOO to support language learning and teaching, this tool offers a rich vein that is still to be tapped, for, as Turbee (1996:2) suggests:

The types of activities that can be created for a MOO are limited only by the imaginations of the teachers and learners.

References

Brooks, J. G. & Brooks. M. G. (1993). *The case for constructivist classrooms.* Alexandria, VA: Association for Supervision and Curriculum Development.

Burdeau, I. (1997). *Virtual Classrooms, Virtual Schools.* Unpublished dissertation for the MA in Media and Language Learning, University of Brighton, http:// homepage.ntlworld.com/ishmael.burdeau/ma/virtual.htm.

Chun, D. (1998). Using Computer-Assisted Class Discussion to Facilitate the Acquisition of Interactive Competence. In J.Swaffer, S.Romano, P.Markley & K.Arens (Eds.), *Language Learning Online: Theory and Practice in the ESL and L2 Computer Classroom,* 57-80. Austin, Texas: Labyrinth Publications.

Crystal, D. (2001). *Language and the Internet.* Cambridge: Cambridge University Press.

Curtis, P. (1992). *MUDDing: Social Phenomena in Text-Based Virtual Realities,* ftp:/ /ftp.lambda.moo.mud.org/pub/MOO/papers/DIAC92.txt.

Davies, L.B., Shield, L. & Weininger, M.J. (1998). Godzilla can MOO, can you? MOOs for Construction, Collaboration and Community and Research. *The Language Teacher,* 22 (2), 16-21.

Dibbell, J. (1993). A Rape in Cyberspace or How an Evil Clown, a Haitian Trickster Spirit, Two Wizards, and a Cast of Dozens Turned a Database Into a Society. *The Village Voice,* December 21, 36-42.

Donaldson, R. & Kötter, M. (1999a). Language learning in a MOO: creating a transoceanic bilingual community. *Journal of Literary and Linguistic Computing,* 14 (1), 67-76.

Donaldson, R. & Kötter, M. (1999b). Language learning in cyberspace: Teleporting the classroom into the target culture. *Calico,* 16, 531-537.

English, J.A. (1998). MOO-based metacognition: Incorporating online and offline reflection into the writing process. *Kairos Journal,* 3 (1), http://english.ttu.edu/ kairos/3.1/features/english/intro.html.

Fanderclai, T.L. (1995). MUDS in Education: New Environments, New Pedagogies. *Computer-Mediated Communication Magazine,* 2 (1), 8.

Hall, C. (1998). *"Constructing" language at MundoHispano.* Unpublished paper, George Mason University, Fairfax, VA. http://www.angelfire.com/ma/CasaDax/ MHpaper.html.

Harris, L. (1996). Using MOOs to teach composition and literature. *Kairos Journal,* 1 (2), http://english.ttu.edu/kairos/1.2/coverWeb/Harris/contents.htm.

Haynes, C. & Holmevik, J.R. (1996). Lingua unlimited: enhancing pedagogical reality with MOOs, *Kairos Journal,* 1 (2), http://english.ttu.edu/kairos/1.2/coverWeb/ HandH/start.html.

Hewer, S. & Shield, L. (2001). Online Communities: Interactive Oral Work at a Distance. In T.Atkinson (Ed.), *Reflections on computers and language learning,* 53-62. UK: CILT Reflections Series.

Hudson, J.M. & Bruckman, A.S. (2002). IRC Français: the Creation of an Internet-Based SLA Community. *Computer Assisted Language Learning,* 15 (2), 109-134.

Jordan-Henley, J. & Maid, B.M. (1995). MOOving along the Information Superhighway: Writing Centers in Cyberspace. *Writing Lab Newsletter,* 19 (5), 1-6.

Keenan, C. (1996). Wading through the MUD: the process of becoming M** literate. *Kairos Journal,* 1 (2), http://english.ttu.edu/kairos/1.2/coverWeb/Keenan/kairosm3.html.

Kolko, B. (1998). Bodies in Place: Real Politics, Real Pedagogy, and Virtual Space. In C.Haynes & J.R.Holmevik (Eds.), *High Wired: On the Design, Use, and Theory of Educational MOOs*, 253-265. Ann Arbor: University of Michigan Press.

Kötter, M. (2001). MOOrituri te salutant? Language Learning through MOO-Based Synchronous Exchanges between Learner Tandems. *Computer Assisted Language Learning,* 14, 289-304.

Lasarenko, J. (1996, Lecture Commentary:1). So you wanna MOO? *Kairos Journal,* 1 (2), http://english.ttu.edu/kairos/1.2/coverWeb/Lasarenko/MOO1.htm.

Lee, S., Groves, P., Stephens, C. & Armitage, S. (1999). *Guide to online teaching: existing tools and projects. MUDs, MOOs, WOOs and IRC*, http://www.jisc.ac.uk/jtap/htm/jtap-028.html.

Levy, M. (2000). Scope, goals and methods in CALL research: questions of coherence and autonomy. *ReCALL,* 12 (2), 170-195.

Marvin, L.E. (1995). Spoof, spam, lurk and lag: The aesthetics of text-based virtual realities. *Journal of Computer-Mediated Communication*, 1 (2), http://jcmc.huji.ac.il/vol1/issue2/vol1no2.html.

Oren, A. (1996). MOOing is more than writing. *Kairos Journal*, 1 (2), http://english.ttu.edu/kairos/1.2/coverWeb/avigail.html.

Papert, S. (1993). *The children's machine: Rethinking school in the age of the computer*. New York: Basic Books.

Parks, M.R., & Roberts, L.D. (1998). 'Making MOOsic': The development of personal relationships on line and a comparison to their offline counterparts. *Journal of Social and Personal Relationships,* 15 (4), 517-537.

Peterson, M. (2001). MOOs and Second Language Acquisition: Towards a Rationale for MOO-based Learning. *Computer Assisted Language Learning*, 14 (5), 443-459.

Rein, R. (2002, A-C:5). *Rachel's Super MOO List*, http://cinemaspace.berkeley.edu/~rachel/moolist/.

Schwienhorst, K. (1997). Virtual environments and synchronous communication: collaborative language learning in object-oriented multiple-user domains (MOOs). In D. Little & B. Voss (Eds.), *Language Centres: Planning for the New Millenium*, 126-144. Plymouth: CERCLES.

Schwienhorst, K. (1998a). Matching pedagogy and technology – Tandem learning and learner autonomy in online virtual language environments. In R.Soetaert, E.De Man & G.Van Belle (Eds.), *Language Teaching Online,*115-127. Ghent: University of Ghent.

Schwienhorst, K. (1998b). *Co-constructing learning environments and learner identities – language learning in virtual reality*. Paper presented at the ED-Media/ED-Telecom Conference, Freiburg, Germany, http://www.tcd.ie/CLCS/assistants/kschwien/Publications/coconstruct.htm.

Schwienhorst, K. (1998c). The 'third place' – virtual reality applications for second language learning. *ReCALL,* 10, 118-126.

Schwienhorst, K. (2001). Evaluating Tandem Learning in the MOO: Discourse Repair Strategies in a Bilingual Internet Project. *Computer Assisted Language Learning*, 15 (2), 135-145.

Schwienhorst, K. (2002). The State of VR: A Meta-Analysis of Virtual Reality Tools in Second Language Acquisition. *Computer Assisted Language Learning*, 15 (3), 221-239.

Shield, L. & Hassan, X.P. (2002). Simulation globale in MOO: designing a framework to define the architecture of anarchitecture. Paper presented at *EuroCALL 2002: Networked Language Learning – a link missing?* Jyväskylä, Finland.

Shield, L. & Weininger, M.J. (1999). Collaboration in a Virtual World – groupwork and the distance language learner. In R.Debski & M.Levy (Eds.), *WorldCALL: Global Perspectives on Computer Assisted Language Learning*, 99-116. Amsterdam: Swets & Zeitlinger.

Shield, L. & Weininger, M.J. (2002a). Directly speaking: the nature of real-time MOO discourse in the light of its role in language learning. Paper presented at *EuroCALL 2002: Networked Language Learning – a link missing?* Jyväskylä, Finland.

Shield, L. & Weininger, M.J. (2002b). Written speech or spoken text? An examination of linguistic output in the MOO environment. Paper presented at *UNTELE 2002: From tool to content or from content to tool in foreign language teaching and learning: which pedagogical rationale to adopt?* Université de Technologie de Compiègne, France.

Shield, L. & Weininger, M.J. (2002c). Promoting Oral Production in a Written Channel: an investigation of learner language in MOO. In J.Colpaert, W.Decoo, M.Simons, M. & S.van Bueren (Eds.), *CALL Professionals and the future of CALL research. Proceedings of the Tenth International CALL Conference*, 201-210. Antwerp: Universiteit Antwerpen.

Shield, L., Weininger, M.J. & Davies, L.B (1999). MOOing in L2: Constructivism and Developing Learner Autonomy for Technology-Enhanced Language Learning. *C@lling Japan*, 8 (3), http://jaltcall.org/cjo/10_99/mooin.htm.

Shield, L., Weininger, M.J. & Davies, L.B. (2000). MOO mythsters, misses and masters: misperceptions, misrepresentations, mistakes, misunderstandings, mastering management. *CALL Report,* 16, 32-40. UK: CILT.

Shield, L., Weininger, M.J. & Davies, L.B. (2001a). A task-based approach to using MOO for collaborative language learning. In K.Cameron (Ed.), *CALL and the Learning Community*, 391-401. Exeter: Elm Bank Publications.

Shield, L., Weininger, M.J. & Davies, L.B. (2001b). MOOtual Muses: shared creation rightfully wrought. In T.Atkinson (Ed.) *Reflections on computers and language learning*, 72-79. UK: CILT Reflections Series.

Shneiderman, B. (1997). Relate-Create-Donate: A teaching/learning philosophy for the cyber-generation. *Computers & Education,* 31 (1), 25-39.

Thorne, S. (1996). *More about MOOs*, http://central.itp.berkeley.edu/~thorne/MOO.html.

truna aka j.turner (1998). *A walk on the ICE*, http://www.fed.qut.edu.au/tesol/cmc/emu/ice.html.

truna aka j. turner (2001). Worlds of words: Tales for language teachers. In U.Felix (Ed.), *Beyond Babel: Language Learning Online,* 163-186. Melbourne: Language Australia.

Turbee, L. (1995). What can we do in a MOO?: Suggestions for Language Teachers. In Mark Warschauer (Ed.), *Virtual Connections*, 235-238. Manoa, HI: Second Language Teaching and Curriculum Center, University of Hawaii at Manoa.

Turbee, L. (1996). *MOOing in a foreign language: How, why and who?* Paper presented at the Information Technology Education Connection's International Virtual Conference/Exhibition on Schooling and the Information Superhighway, June 3-6, http://Web.syr.edu/~lmturbee/itechtm.html.

Turbee, L. (1997). *Educational MOO: Text-based Virtual Reality for Learning in Community*. Syracuse, NY: ERIC Clearinghouse on Information & Technology, http://ericit.org/digests/EDO-IR-1997-01.shtml.

Vygotsky, L. (1978). *Mind in society: The development of higher psychological processes*. Cambridge, MA: Harvard University Press.

Weininger, M.J. & Shield, L. (2001). Orality in MOO: Rehearsing Speech in Text. A preliminary study. In K.Cameron (Ed.), *CALL – The challenge of change*, 89-96. Exeter: Elm Bank Publications.

MOOs

Dreistadt MOO – http://cmc.uib.no/dreistadt/informationen2.html, EnCore Web Interface: http://cmc.uib.no:7001/

Esc-Pau MOO – http://moo.esc-pau.fr:7000, Telnet MOO client: moo.esc-pau.fr:7777.

GrassRoots MOO – http://www.enabling.org/grassroots, http://www.enabling.org/grassroots/java-new/, Telnet MOO client: enabling.org:8888.

Le MOOFrançais – http://www.umsl.edu/~moosproj/moofrancais.html, Telnet MOO client: admiral.umsl.edu:7777.

Le MOOlin Rouge – http://cmc.uib.no:9000/.

Mundo Hispano – http://www.umsl.edu/~moosproj/mundo.html, Telnet MOO Client: admiral.umsl.edu:8888.

Schmooze Universty MOO – http://schmooze.hunter.cuny.edu:8888, Cup-o-Mud Web Interface: http://schmooze.hunter.cuny.edu, http://schmooze.hunter.cuny.edu:9000, Telnet MOO client: schmooze.hunter.cuny.edu:8888.

All websites cited in this chapter were verified on 28.08.2002.

Acknowledgements

Thank you to all the participants in the various MOO projects with which I've been involved over the years for their generosity in allowing me to carry out research on their words and actions. In particular, I'd like to thank the administrators at *Schmooze University MOO*, *GrassRoots MOO* and *Esc-Pau MOO* for supporting those projects.

Markus J. Weininger, truna aka j.turner and Xavière Hassan read the drafts and made suggestions and corrections – thanks to them for their patience; any remaining errors are mine alone.

None of this work would have been possible without the support of my good friends and co-researchers Markus J. Weininger, truna aka j.turner and Larry Davies; their imagination, encouragement and most of all their friendship has been invaluable – they are the proof that MOO works! To them go the biggest thanks of all.

7

Virtual worlds as arenas for language learning

Patrik Svensson, Umeå University, Sweden

Introduction

Traditionally, technology seems to have had a rather clear function in language learning: for example to practise grammar, provide information about target countries, make long-distance communication possible, bring in the outside world and access digital corpora. The computer can thus be seen as a versatile and widely applicable tool, and the learner could be described using Levi-Strauss's concept of the *bricoleur* who 'is adept at performing a large number of diverse tasks' (1966:17). This paradigm can be contrasted with a view of technology as an automaton (Skinner 1968), according to which technology plays an important role in giving feedback and reinforcing material that is taught. In language learning, drill-and-practice applications provide good examples of this latter way of using computers. Typically, these programs have a strong sense of what is right and wrong, which might not always correspond to how language is used in the real world, but they do of course, have a clear pedagogical function. In this chapter, we will be concerned with technology used in language learning, not primarily as a tool or an automat, but as an arena for constructivist learning (see Svensson and Ågren 1999). A three-year project, the Virtual Wedding Project (VW Project), carried out at the Department of Modern Languages and HUMlab at Umeå University will serve as a case study. In this recently finished project, advanced students of English constructed collaborative assignments in the target language, contexualised in a graphical virtual environment in which they acted as virtual construction workers, hypertext authors and community builders.

One of the prime driving forces behind the Internet and new technologies is undoubtedly the eagerness to communicate and to spend time in different types of computer-generated environments. We spend more and more time on communicating by email, playing computer games, and experiencing different kinds of virtual environments. The Swedish community platform Lunarstorm attracts more than a million people every month (mainly young people), and it is clear that there is an enormous feeling of community and shared space in an environment that typically has at least 20,000 people logged on simultaneously. With the help of mobile devices the sense of community is easily maintained even away from the computer screen (Rheingold 2001). The computer game industry generates more annual revenue than the global film industry (Berger 2002), and game environments are becoming increasingly shared and distributed. Graphical virtual environments such as *Active Worlds* attract hundreds of thousands of people, and computer-generated satellite photos of these worlds show us virtual urban geographies that have a history and life. It makes sense to think about how such resources can be used creatively in learning contexts, and how we can use them to support the kinds of language learning we would like to make possible.

Sometimes it appears that technology, software and pedagogy make technicians and teachers the producers of digital content, while students are users rather than producers. In many cases, this must probably be so, but in the following, it will be argued that it is necessary to let students create streamed media, hypertext and virtual worlds to encourage a more profound understanding of the media, stimulate motivation, and to use technology as a means whereby students can construct knowledge themselves. These creative aspects of the new media have a great potential impact on language learning, and one possible development is to allow more room for creative and artistic means of expression in an educational context that is still largely textual and oral. This context also involves other kinds of places than classrooms and educational institutions. The idea of students creating, sharing and developing such arenas might not be new, but is certainly exciting and challenging, offering the opportunity to interact in the target language in a way that is both authentic as to context and meaningful for the interlocutors, so providing a motivation that may be absent from the traditional language learning context.

Virtual worlds

EVE (English Virtual Environment) is the graphical virtual space which has been used for the VW Project which will be described in more detail later. It encompasses some 360,000 square meters, and it is interesting to note that virtual worlds can indeed be measured as we measure non-virtual space. There are more similarities between virtual worlds and non-virtual worlds than we might expect. The possibility of creating worlds with different physics and new ways of visualising information is relatively rarely exploited. Much of what we find in virtual environments such as *Active Worlds* bears a remarkable similarity to a random American city, a business setting or a university classroom. It is clear that we are to some extent concerned with recreating reality in the computer. There are many kinds of realities, however, and the relationship between human perception and virtual reality is neither simple nor fully explored. Nevertheless, in most learning contexts we are not really interested in creating highly detailed replicas of real-world entities. This is very costly, time-consuming and does not really appear to be an overly creative use of the medium. Most language teachers would probably prefer to send their students to the target country rather than recreate it in the computer at much greater expense. Recreating physical spaces might be called for in many contexts, but often it seems a waste of time to recreate university buildings in great detail in virtual space. Somehow, the university building is easily confused with the concept of what a university is, and there is some risk that such a virtual university will actually reinforce traditional learning (through the traditional setting) rather than promote changes and accommodate new learning paradigms. Furthermore, a great deal of detail might not encourage creative interpretation but instead present us with something that is ready-served and that does not necessarily invite interpretation and exploration (see McCloud 1993:30-31 for an interesting parallel in relation to comics).

There are many different types of virtual reality (VR) ranging from highly immersive environments with VR-helmets, spatial sound and force feedback gloves to graphical onscreen environments (sometimes called 2 $1/2$ D environments) and textual environments. Text-based MUDs (Multi-User Dungeons) and MOOs (MUD Object Oriented) have been used in educational contexts from the early nineties (Cherny 1999); at least on a limited scale (see Shield this volume). One of the key attractions of these environments is that they open up the classroom to a larger world, and that they facilitate communication and user-changeable shared spaces. The world is presented through text and communication is text-based. There is an inherent spatiality in the world, and as a user, one moves about through the various locations that make up the world. Users generally take on another role or persona in world, and text worlds have proved to have a strong immersive power. Inhabitants of these worlds become involved and sometimes even obsessive (Turkle 1995). In language learning, role-playing and simulation play an important part, and virtual places can provide very useful arenas for such role-playing. Paradoxically, virtual simulations often turn out more 'real' than ones that are carried out in the classroom. An example from our own work would be several European language simulations: for instance the *Achill Project,* where students from a range of European countries have co-operated through video conferencing, email and web pages. Job applications and job interviews, for example, proved to be much more realistic in this kind of setting than in the traditional classroom. The key element, of course, is meeting other real people ('playing' themselves or having alternate personae) and working collaboratively with remote participants. As the example above indicates, this kind of role-playing does not need to take place in virtual worlds. It can also be structured by web pages, email and other means, but virtual worlds have the potential to create a place and a unified spatial interface.

Non-physical spaces for language learning

A key question here is what is really entailed in carrying out learning activities or any kind of activity in a basically non-physical space. In line with the idea of regarding technology as a tool with very specific functions, there is a tendency to regard graphical virtual worlds, or just spending a great deal of time on computers as a waste of time. It is important to acknowledge that being social or thinking online is just as valuable as carrying out the corresponding activities in the real world. In particular, we need to realise that online experiences are not necessarily secondary to real-world experiences or just non-real. As Markham (1998:120) points out, experiences that are experienced cannot be 'not real'. Language that is used online (whether chat, email, SMS or voice) is not peripheral, and as language professionals we need to work with this kind of language, and introduce our students to it. There is also a growing literature on various theoretical aspects of virtuality (Hayles 1999, Lunenfeld 2000) which would seem to be relevant for making extensive use of virtual environments in education. For instance, the role of the body is one important factor. When science fiction writer William Gibson and others pictured cyberspace in the 80s they often did so from a rather dualistic perspective in the sense that the body is not very important. What counts is information transmittable through digital networks, and the human soul ultimately turns into data. Of course, re-embodiment frequently takes place in cyberspace when a human user takes on a character (a so-called avatar) in a virtual world. What does it mean for learning that in a sense the learner leaves his or her real body behind? From a cognitive perspective (Lakoff and Johnson 1999) it might be argued that everything we do is based on the fact that we are embodied beings. Knowledge is not disembodied or objective, but constructed and experienced. We use our hands, vision, brains etc. to interact with the technology and to make sense of what the computer presents to us. Graphical environments where users manoeuvre human-like representations around are based on the fact that we function in certain ways, and this is one of the reasons why virtual worlds tend to resemble real worlds so closely. Bodies and spaces play important roles in almost any communicative situation, and their virtual counterparts can be useful in language learning, not least because of the dynamic and distributed nature of virtual worlds. Another important aspect is the natural co-evolution between embodied beings and their environment, which shapes and constrains both. We will describe the virtual environment under discussion in detail later (also in pictures), but first we will present a very brief overview of graphical virtual worlds.

Graphical virtual environments (see Damer 1998 for a good introduction) tend to look like computer games with a 3D space and user controllable characters moving about in that space. In *Active Worlds* one can either choose to see one's own avatar (body) or not. Avatars are mobile (and have the ability to fly), may express certain gestures, and one avatar may be exchanged for another one (for example a man in a suit may suddenly turn into a hippie woman or even a zeppelin). The students who have worked in our project have acquired considerable skill in shifting between perspectives, moving between spots, changing avatars, and organising in-world meetings. This kind of knowledge allows them to work with language, communication and collaborative building in the world. As was pointed out earlier, most of the worlds tend to be rather realistic. The perspective used in *Active Worlds* has a tendency to make the avatars rather small and it might be argued that the exteriors in some sense are given high visual prominence. DeVarco (2002) distinguishes between avatar-centred and place-centred environments, and while there is considerable variety in *Active Worlds*, it seems as if the platform itself is predominantly place centred. If we want to recreate a portion of the country where a language is spoken or a specific setting such as a court, we might prefer a place-centred platform. If we, on the other hand, focus more directly on communication, it might be more suitable to use an avatar-centred platform. It is important to realise that the various platforms available not only have advantages and disadvantages, but also to a large extent impose a structure on learning activities carried out in that environment. While the possibility of building your own world easily is important in *Active Worlds* it may not be the same in other environments. The platform *Traveler* is largely avatar-centred rather than space-centred, and here the avatars are typically represented by enormous heads. The surrounding landscape is usually abstract and rarely detailed, and this combination is often stunning. We know from experience and research (Liggett 1974, Massaro 1998) how important the face is to human communication, and how much information it conveys. A key difference between the two platforms described above is that while direct communication in *Active Worlds* is text-based, *Traveler* depends on live audio communication, and this is one reason for using prominent heads in the latter. Lip-synching technology creates an impression of someone actually talking. Another difference is that *Active Worlds* provides an integrated browser window which facilitates powerful integration between the world and hypertext. *Traveler* has no equivalent facility. Yet another platform, Adobe's *Atmosphere*, which is still under development, provides high-quality graphics (much less cartoonish than *Active Worlds*) and real-time physics, while building facilities so far are restricted to a stand-alone application. What is important here is not the specific platforms, but rather some of the parameters that might have relevance in a language learning context: building features, place-centrism and avatar-centrism, communication through text or audio (or both), level of graphical detail, and possible integration of a browser. Needless to say, other relevant factors include price, ease of administration and user friendliness.

Social worlds

Most successful platforms for creating virtual worlds, regardless of whether they are mainly place or avatar centred, depend heavily on the social function of the world. While it may be exciting to inspect replicas of real world places, shop in virtual shops, or visualise a planned suburban area, most users of online worlds are there to meet other people. This means that any attempt to partly replace the classroom with a virtual place has the potential to make the learning situation more open and bring in other people, most importantly native speakers of the target language. Predictably, these environments readily lend themselves to various types of collaborative and constructivist work. Students share a world which they build themselves and where they can work together as well as individually. There are several advantages to this type of collaborative arena. First, there is a distinct place (albeit virtual) which is tied to the collaborative work and which can be explored and developed by the students. It does not matter whether participating students are in the same physical location. Secondly, this arena reflects both the process of learning and the 'result' of learning. There might not be a product in the traditional sense (such as a critical analysis of a literary text) but rather an environment that a group of students have co-developed which in itself also has a history. This environment is a shareable graphical space present in a computer, and student groups have their own part of the world to develop. Using this space, they create graphical representations and link them to academic texts. The world itself remains, and the work of previous student groups might be present in the same environment. Thirdly, the openness of most virtual environments is a strong motivating factor. Students feel that they have a real audience out there. They might also meet outsiders who are interested in the work they are doing, and participants from other places may give presentations or just come by to chat. In this way, language students are encouraged to express themselves and to communicate in an environment that is not controlled by the teacher. If several groups are working on collaborative projects, a shared environment might give rise to more between-group interaction. This is particularly useful if native speakers are integrated as participants since they provide authentic genre-specific discourse models with which a non-native teacher may not be familiar.

The power and myth of virtual worlds and virtual reality (the early stages are well portrayed in Rheingold 1991), though, are not reliant on any specific feature such as collaborative or social aspects. Davis (1998:248) says that 'the belief that VR constructs a world, a simulacrum powerful enough to temporarily overwrite our material one, has been embraced as an article of faith by the technology's fans and detractors alike'. This world, whether we call it virtual reality or cyberspace, deserves both healthy scepticism and intense exploration. There is certainly something enticing about immersing oneself in virtual space, whether textual or graphical, as is evident when Anders (1999:92) describes the fantasy states that a text world can give us as 'a form of communal dream or memory palace defined by words – yet experienced as space'. The point here is that virtual reality and cyberspace do not only provide arenas for learning, but also give us new subject-matter; not least in the liberal arts. As language teachers, we might want to explore narrative structure in computer games, how language changes in cyberspace, how culture and identity are affected by the new media, or how the second wave of hypertext novels are changing the notions of author and reader. Here there is also room for far-reaching collaboration between technical subjects and liberal arts subjects. This development seems logical if we think about technology not only as a tool but also as a place and an arena.

Challenging a text-based paradigm

Before describing the VW Project in more detail, it should be acknowledged that it was preceded and has been supplanted by some related projects, and experience from a Hypertext Project and several European large-scale collaborative projects has fed into the project. For about five years now, teacher trainees doing their third semester of English at the department write degree assignments as hypertext essays rather than as traditional essays. These students go to England for ten weeks, come back to Sweden, and are then required to write this piece of work. They are a suitable group with whom to try out hypertext presentations as they need to work with new technology to prepare for their future careers and because they bring with them a wide range of materials from England. The assignments typically focus on aspects of culture and life in Britain and Sweden. The students present the finished work to fellow students and to teachers who have been involved, and this presentation has to be done with the help of a computer and a projector. Each student has to present their project in front of a digital video camera, and these short clips are collected on a CD together with the rest of the work. One thing we have learnt is that it is very satisfying to the students to have a physical artefact that represents the work they have done; something that can be shown to a future employer for instance.

One aspect of the Hypertext Project that had an important bearing on the VW Project is the process of going from 'paper' assignments to hypertext ones. As we observed earlier, there is traditionally a strong textual paradigm, and as Felix (this volume) points out, there are few studies that have analysed the role of graphic elements in language learning. This paradigm affects both teachers and students, and the transition is rarely trivial. It is certainly a matter of threshold, and while many students are initially hesitant in the Hypertext Project, they almost exclusively end up being rather enthusiastic hypertext workers. Furthermore, in evaluations they say that they would now be willing to take part in projects which have to do with information technology when they start their work as language teachers. It is remarkable how technology is often seen as pure magic and something that others do, and how simple it really is to break that spell. Here educators, not least language professionals, have a very important role to play. We need to allow our students to become confident users of new technology and creators of new media. Ultimately it is an issue of silicon literacy, and 'the ways in which meanings are made within these new communication systems' (Snyder 2002:3).

Turning back to the transition from paper to hypertext, one risk teachers may face is that they do not state explicitly that they expect the students to work with a new medium, and that paper is no longer the primary one. In the Hypertext Project, the first rounds were negatively affected by the fact that we wanted the students to hand in a printed version so that we in some ways signalled that the paper version was still important. Creating a hypertext essay is not about taking a Word file and simply converting it to a web page. We need to work with the medium, and think about how it can be used most suitably for our needs in the new context. Every medium has its conventions, and we need to be aware of these.

The Virtual Wedding Project

Introducing a virtual environment in addition to a hypertext element represented the next step, and meant that we were able to bring together the creative and analytical aspects of hypertext with the collaborative and world-building aspects of virtual worlds. In particular, it meant that it made possible a unified interface to a rich multimodal and distributed platform with a visual environment and rich hypertext. We found that many of the issues we encountered in the new project were similar, and we were continually guided by the following questions. What does it mean and entail to work in a different medium? How can we facilitate creative use of a new medium? How is it possible to stimulate early 'immersion' into new media? How can hypertext and virtual construction be assessed (as opposed to traditional essays)? How does analytical work in hypertext and virtual space relate to traditional analytical writing (on paper)? What role do the visual and multimodal elements play?

The Virtual Wedding Project was financed by the Council for the Renewal of Higher Education in Sweden for three years. The full title of the project was 'Cultural simulation: Virtual weddings and a real wedding of linguistics, literature and cultural studies', and as the title indicates, one of the basic ideas behind the project was to bring together the sub-disciplines that language subjects at Swedish universities typically incorporate in their programs. In addition, the project aimed to develop analytical skills, explore new technology, and open up the classroom to a wider, active audience. In the project, students built a graphical online world and linked physical constructions in the world to a rich set of web pages, some of which represent hypertext versions of the sorts of academic essays described below. They worked extensively both in the world and outside the world, and the project can probably be said to be process-oriented rather than product-oriented. There was a final product, the graphical environment with all the linked web pages, but this world was not an endpoint and the graphical constructs can be seen as a reflection of the learning process.

The project involved advanced students of English, and corresponds to a major academic assignment that is the equivalent of ten weeks' work (half a semester). Traditionally, these essays have a very specific format (footnotes, bibliography, structure etc.), are specific to the various subdisciplines, and are largely individual projects. Representative examples from 2001 include: 'A struggle for power – a reading of Margaret Atwood's Alias Grace' (literature); 'Image is everything. A Study of body, sexuality, values and influence in two music videos' (cultural studies); and 'Place-names in New South Wales, Australia' (linguistics). They typically encompass 20-25 pages, and are intended to teach students about academic and analytical writing, and to prepare them for the fourth term assigment. The essays are also predominantly textual. Out of the 22 assignments that received a pass grade at the department between January and May 2001, three contain graphical elements (other than tables and in one case, a bar chart). Two of these essays deal with the language of advertisements (pictures of advertisements), and one linguistics essay contains a few dialect maps. Even when the subject-matter (music videos and images) would seem to need visual elements, we rarely find any kind of images. The main point here is not that textual representation is not good enough but rather that the conventional paradigm is extremely text-oriented.

The idea of the project was to offer an alternative to these traditional assignments, and to encourage collaborative and non-traditional work around a central theme in a virtual environment, open to the outside world. The theme had to be rich in the sense that it had a bearing on linguistics, literature and cultural studies. In Sweden, the tradition is for language departments to offer courses in linguistics and literature as well as proficiency and communication skills, and over the last decade, cultural studies has become a sub-discipline of its own at some universities. However, it is remarkable that even though these sub-disciplines co-exist at the same departments and share the same students and programs, there is a great deal of compartmentalisation. In the project, cultural studies was seen as a bridge that could bring linguistics, literature and communication together. It should be pointed out, however, that the approach used in the project is not restricted to this particular academic setting. It could easily be used in more communicative-focused settings, and also in other disciplines. The themes that have been employed over the three years are weddings, the city, and monstrosity. Weddings, the initial theme, was chosen because it was an approachable topic which could serve as an entry-point to scholarly analyses, and because this theme lent itself to the creating of a culminating happening in the virtual world (a virtual wedding).

Students were offered the chance to participate in the project as the equivalent of the ten-credit assignment described earlier, and they had a free choice between writing a traditional essay and taking part in the project. The themes used were chosen after the previous round had been finished. There were particular reasons for each choice, but more generally, we wanted themes that bridged the sub-disciplines, that were topical and attracted current academic interest. We also found that some of the themes provided a valuable link between undergraduate education and on-going research. In total, 22 students signed up for the three runs. The last two rounds each attracted almost half of the students in the program. The majority of the students were women and very few had any advanced computer experience. Most of them, however, had used word processors and web browsers. In the project, they were given a brief introduction to web page creation and to using the virtual environment software (two three-hour sessions). In the latest round, the media engineering program contributed two media engineer students who worked with the language students as a part of their course in providing computer support. Nevertheless the student participants themselves relatively soon acquired a reasonable high level of expertise, and it was difficult for the media engineers to keep ahead. In all the iterations, the students split up into teams (typically three to four members in each team), and it seems that this setup allowed the students to learn the technology from each other – both in their own team and from the other teams.

In all three iterations we have made use of HUMlab which is a state-of-the-art humanities technology laboratory at Umeå University. HUMlab has been designed to be a meeting place between the humanities, culture and art, and new technology and new media. It makes a mixed array of technology available, including 3D projection screens and powerful computers, and the setting is very non-traditional with sofas, an aquarium, books and colours. It is a regular occurrence that many activities take place in the laboratory simultaneously. While we have not needed all of the equipment (in fact, the demands on hardware are rather low), the students have been exposed to new technology, and the whole setting with many technology students, visitors and events has had a very positive effect on the project. The laboratory has also hosted the final presentations. These presentations have been mainly virtual, but there has also been a local crowd present, and web cameras have broadcast the laboratory setting to virtual visitors from all over the world (onto big posters in the world). The two final presentations have attracted about thirty in-world people.

Visiting EVE

The central arena of the project has been (and still is) the aforementioned virtual world EVE, and to make things a little more concrete, an example screenshot from the world is displayed in Figure 1.

Figure 1. Active Worlds screenshot with world, chat, and browser windows

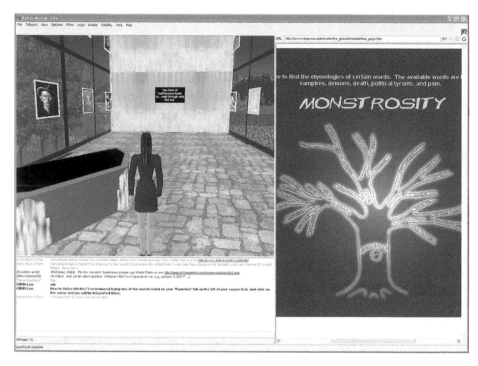

Here we see the *Active Worlds* program running the world EVE, and there is one avatar present (the author in this case). We could have chosen a first person perspective in which case we would have seen no avatar at all. Of course, if the world had been busy and populated, there might have been many other avatars around. The particular visual environment presented above is an example of student work on monstrosity. The screen is split into three sections: the world window, a chat window, and a web browser window. The world window gives us the visuals, and when we move about in the world this window will change. When other people come close to us we will see them here etc. The chat window provides facilities for chat conversation, near-synchronous text-based conversation, and the web window shows us the web page associated with this particular part of the world. Moving about in the world and clicking on objects often triggers changes in the web window. Walking to a student group's or an individual student's space will typically call up a group web page or a student page. This facility is very important as it allows integration between the shared visual space and any number of web pages. As we would expect, the web browser offers us an opportunity to link in not only basic web pages (text, images) but also streamed media, sound and other resources. It is also possible to create links from the web window directly to specific spots in the world (web teleports). Through using different kinds of building blocks, users can create new artefacts in the world. There are large numbers of objects and avatars on the Internet, and if there is a need, 3D modelling software can be used to create new objects and avatars. In the VW Project, we tried not to let the graphical detail take up too much time but encouraged the creative use of building blocks that are already available. In some cases, students thought that the inventory was somewhat limited for their specific theme.

When the project was initiated, EVE was an empty world, and we created a tower at the very centre to have at least something in the world. The idea was that as the project progressed, the world would grow, and eventually, watching the world from that tower would give the viewer a panorama of projects, structures and activities spread out all over the world. We had not quite realised that if users need to get a good overview, they usually just fly high up in the air! With the exception of the tower, the world as it stands today is thus purely a result of student work.

Ascending Bloom's taxonomy

One of the original ideas of the VW Project was to use Bloom's taxonomy as a starting point, and in this figure, the basic action structure of the project is shown.

Figure 2. Project workflow in relation to Bloom's taxonomy.

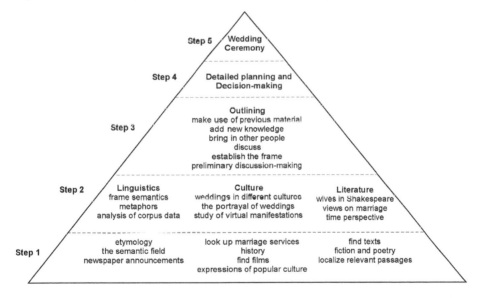

Bloom's well-known taxonomy of cognitive objectives (Gage and Berliner 1998) ranges from knowledge, comprehension, application, analysis, synthesis to evaluation. There is no one-to-one relationship with the steps in the pyramid, but the general idea was to go from material-oriented tasks in the three sub-disciplines to more analytical tasks, and to a greater degree of collaboration, negotiation, and synthesis. Language skills come into all these areas: communication, negotiation, writing, virtual and real meetings, writing invitation cards, explaining the world to visitors and so on. In his later work, Bloom also suggested objectives in the affective and the psychomotor domains, and in relation to constructivism, it would seem tenable not to focus solely on cognitive objectives (see Gagnon and Collay 1996). Learning is about experiencing and engaging in social exchanges as well as working with material and reflecting. As the project has developed, more emphasis has been put on thematic aspects (concepts), external participation, in-world construction and creativity. Furthermore, the sequential nature of the taxonomy represented in the pyramid has become less rigid, and indeed, it would seem that a constructivist approach might not be fully compatible with the clearly separated levels in Bloom's taxonomy. This does not mean that Bloom's taxonomy is not very useful, not least for moving 'beyond recitation of information as the goal of education' (Gagnon & Collay 1996:5).

One general problem in modern university education seems to be the difficulty of reaching the higher levels in Bloom's taxonomy, and teachers often describe it in terms of students who are not used to working analytically. At the Department of Modern Languages/English which has provided the academic setting for the project, the pass rate for third term assignments generally seems to be about fifty percent. Fifty-four percent of the non-project students had finished their 2001 spring semester essays more than a year later. The pass ratio for the VW project for the same semester was 100%. Out of a total of 23 students, 10 students had chosen to work in the project. This, of course, indicates that the VW students were highly motivated as self-selected participants. From close observation we feel, however, that the collaborative character of the project, the motivation of the medium, and the constructivist work process contributed a great deal to the positive outcomes. Generally speaking, the students enjoy the project, and the prospects for future iterations are very good. Expectedly, there have also been many stumbling blocks along the way and problems that would not occur in the traditional essay-writing paradigm. For example, students have to adapt to the medium and have to be willing to explore a virtual environment. We have gradually tried to adopt a 'plunge-in' approach, whereby students get to meet people and do building in the world at a very early stage in the project. For further evaluative details the reader is directed to the project website where an external project evaluation is available.

Making the transition

The fact that the VW Project is in many ways very different from regular assignments makes 'translations' between the two types of academic work difficult. Moreover, the holistic nature of the project precludes simple analyses of small bits. It is not possible to look at one small section of the world, or one or two hypertext assignments. Virtual worlds are not 'read' in a linear fashion, and it is difficult to know where to start and where to end. As supervisors, we have graded the students' work, and found that analytically, and from the point of view of content and use of the media, they have generally presented solid academic work at high levels of linguistic proficiency. It should be pointed out that they have created a great deal of hypertext content, and in fact, most of them have 'produced' as much text as the 20-25 pages of a regular essay. Despite this fact, students are very positive about having participated in the project. In grading student work, we have looked at the overall analytical enterprise as well as at the hypertexts, the in-world representation and the teamwork. We do not expect students to master the technology or design high-tech web pages, but we do expect them to work creatively with the medium through experimenting and exploring. The project has resulted in a wide range of hypertext assignments and places, and some of them are clearly more creative than others. In some cases, it is obvious that the medium makes things possible that would not have been possible before. For instance, there is a piece on Black Vernacular English where the student has created a useful and analytical web structure as well as a strong multimedia component with sound recordings. All this is integrated in a set of web pages and in the world. In Appendix 1, a hypertext work by the same student is shown. It should be borne in mind that when running the appropriate software, the browser window is part of a larger context with the graphical representation of the world and a chat window. What is shown here is the starting page for a set of web pages related to attitudes to language in an urban context.

In order to further illustrate what working in a non-traditional medium may involve, we will end this section by looking at a very specific part of the project: the final presentations. The screenshot in Figure 3 is from the 2001 presentation, and it should be quite clear that the theme of this year was neither weddings nor monstrosity. The presentations take place in the virtual world itself.

Figure 3. Project presentation in Active Worlds (world window)

There were three teams in the city theme iteration of the project, and what we see in this image is the end of one of the team presentations (Team North). We are located in the central city square, and on top of the pillars we see images of some of the team members (there is also a third pillar not visible here). One way of gaining access to the student work is to click on the photos. The team members are visible below their respective pillar (up in the air). There is also a poster here that shows what was going on in the HUMlab at this minute. The audience is mixed, with people from many countries. At least three of them are international authorities on virtual worlds and educational technology who have taken an active part in the project. What we see here is a snapshot and when this event was actually happening, people were moving about in the world and the students were presenting their work, how their city had been constructed, and guided people through the city.

Carrying out a virtual presentation is obviously different from carrying out a real-world presentation. In this particular world, text is used for direct communication. Typically, presentations are relatively short (less than 20 minutes). There are ways in which a virtual world can be designed to work well for presentations for large audiences, and there are ways to get people to move about in the world in a relatively orderly fashion. What is really amazing is that it works very well if the event has been well planned, and this means that we can actually have student presentations in a coherent visual world in front of an audience with people from all over the world. Naturally, language plays a vital part, and the students involved develop a communicative competence that is not restricted to a virtual environment. The excerpt in Figure 4 is an example from the project presentation in 2002 (monster theme). Here we have students communicating with international guests in the world through text. The virtual presentation is just about to start, and the students are taking care of last-minute preparations, and this short passage gives a sense of the anxiety and energy that precede most public presentations.

Figure 4. Getting ready for the presentation (chat window).

```
Emma: hello!
Emma: So, do we lead the guests to Ghould first?
Emma: ghouls
"Mark": Erica, Michel is going to log on, and needs coaching
in turning on tourist whispers.
Erica: Mark.. do you have e-mail handy???
Nick: Mary, you look like a triangle on my screen..
"Mark": I can bring it up in another window, ues ...
"Colin": Hello
Erica: Ok Mark easy enough
Mary: I look fine on mine
Steve: hello, ghouls
Emma: you are a triangle on mine too.
Nick: Ok, sorry.. Maybe I should empty my cache...
Emma: there you are Mary
Peter 80: welcome to our world :)
"Mark": I am logged into my e-mail account ...
Nick: Now you look fine.
Erica: Mark write a quick note to Nev@aw.com and tell him to
tell you the password for your account.
Peter 80: Steve is a black square
"Colin": anyone else here visiting from the University of
Melbourne?
Steve: nope
Mary: I've chnage shape
Emma: Is Dr. Stallworth here?
Nick: Yes, I can see that.
Nick: Hey, Bob.
Emma: Do you think we should wait until 6.
Mary: Did I turn into a rectangle now?
```

```
Emma: that is another 4 minutes
Steve: we probably should
Peter 80: Emma: yeah, let's wait a few more minutes
```

We are not, however, just concerned with communication (such as video conferencing or chatting), but there is also a strong sense of place and other people being present. This sense is partly represented by the integration of the visual environment with the chat environment. The avatars are moving about in a space at the same time as they are communicating and carrying out activities in the world.

Presence is a non-trivial issue in graphical environments, and it is clear that these worlds are not just augmented chat places. For instance, research shows that users have a sense of personal space which is also expressed through the proximity of avatars (Taylor 2002:42). From a communicative point of view, avatar configuration is a way of signalling who is communicating with whom (just like in the real world). In virtual environments such as *Traveler* where there are big headed avatars and real voice, the present author has often felt reluctant to interrupt a circle of talking heads. This is exactly analogous to not wanting to interrupt a group of people taking part in an engrossing discussion. Hopefully, this example gives a better understanding of what working in virtual worlds might involve, and there can be no doubt that the students who have carried out in-world presentations have acquired a unique experience and competence. They have not only built a world, but they are also showing it to the rest of the world.

Constructivism and creativity

Many of the ideas underlying the project harmonise well with the general framework of constructivist learning (see Felix this volume). Naturally, the interrelation between pedagogical ideas and the use of technology is vital, and it is worth pointing out that the technology should facilitate the kind of learning we envision as teachers. The constructivist learning paradigm brings in many of the pedagogical ideas and views that are important to the idea of creating virtual arenas for learning. According to Fosnot (1996:ix), constructivism suggests an approach which 'gives learners the opportunity for concrete, contextually meaningful experience through which they can search for patterns, raise their own questions, and construct their own models, concepts, and strategies.' As articulated by Dewey and others (see Boisvert 1998), there is an emphasis on learning as an active process and as a social activity. Motivation is a key concept, and there is a focus on student activity rather than teacher activity. The body and the senses are important, and constructivism rejects an objectivistic view of knowledge (called the spectator theory of knowledge by Dewey). Hence there are many parallels between this paradigm and the experientalist, 'embodied' approach described earlier in this chapter. Interestingly, recent technological development clearly has the potential to support and enhance constructivist learning as the technology is well adapted to creative work, communication, experiencing and constructing models. Indeed, graphical virtual environments and language learning would seem to make a particularly good testing ground for a view of knowledge as physically and symbolically constructed.

Through looking at some more student work in the VW Project we will now attempt to relate constructivist learning to virtual arenas and the general workflow in the project. In the second iteration of the project, the city theme, we asked students to work with concepts related to the city (and the sub-disciplines) and to represent these in the city. Each group had to come up with a small number of concepts. This work resulted in very lively cities with a number of conceptual focal points (such as gender and materialism). The process of establishing the concepts and motivating them went on for many weeks, and turned out to be a successful way of encouraging in-group collaboration and negotiation. In addition to the virtual space, students worked with hypertext essays linked to the world.

Figure 5. Virtual recycling station (world window).

In Figure 5 we see something that we will find in most cities, but much more rarely in academic work: a recycling station. This is a small bit of the city that Team North built, and the idea is to represent the work that led up to the concepts that this student group eventually chose. Here we find dumped concepts (entertainment, communication, and the city as a trend-setter) and dumped hypertext ideas (the walkman and the friendliness of the stranger). Visitors are encouraged to recycle these ideas, and through clicking on the trashcans, we get to know why they did not work. The rest of the city has both been created to 'give the feeling of a city' (as one of the student puts it in the streamed video clip that presents the city) and to represent the central concepts. For this group, the concepts are diversity, freedom, future visions, and representation. Usually there is a number of hypertext essays linked to every concept, and for instance, the representation concept includes essays on the city in Dickens and Austen, personification of the city in poetry, high versus low in the city, and urban representation in various TV series. The virtual environment brings the concepts and essays together in a spatial domain, and behind the spatial configuration, there is a flexible workflow that structures student work in the world.

The recycling station is an example of student creativity, and creativity and motivation make up important aspects of learning that must not be disregarded (see Debski 1997). New media and a constructivist approach may help to stimulate creative learning and experimenting; not least through establishing social and artistic spaces. Text can certainly be creative and dynamic, especially in dynamic environments such as MUDs, but somehow it seems that visual and auditory means of expressions in distributed spaces have a very strong motivating and creative effect. Recent research focuses on how media can be used creatively in schools. Danielsson (2002:229) reaches the conclusion that 'video and other media can be a vital instrument in young peoples *creation of ideas*'. In order to illustrate some creative use of a multimedia setting and a virtual environment, we will return to the same 'pavilion' we visited earlier.

Figure 6. The path of indifference (world window).

Figure 6 shows the waterfall that can be found at one end of the glass room in the project in which the main theme is 'indifference represented as a monster'. The student creator has used the medium to create an effect and an experience. After having walked through the pavilion, with access to various essays and media, the visitor reaches a waterfall with a sign. It is not possible to see through the water, and we do not know what to expect on the other side. The whole setting with its music is very suggestive, and upon entering the waterfall, the user falls down into a narrow space. Basically the avatar is trapped there, and the waterfall behind is not directly accessible because of the height difference, and in all other directions the user is confronted with high walls plastered with photos of human monstrosity. On a sign there is some text about Plato's gadfly and directions to which keyboard keys will lead to freedom. The building of this space is certainly about exploration and creativity, and it should be pointed out that these students (and many others working with new media) are in fact exploring the extents of the technology and the medium. In the process of exploring and creating arenas such as the one described here, the target language becomes a communicative vehicle through which the students are naturally motivated to express themselves academically, socially and creatively.

Conclusion

We have only started to explore the possibilities of virtual worlds in advanced language learning, and while there obviously is a long way to go, it seems clear that they have a great deal to offer. Language learning is about language, immersion in other cultures, communication, media, intercultural meetings and role-play, and virtual arenas supply us with a place where all these can come together. It has been argued that we need to be open to exploring these worlds as well as letting students do much of the exploration themselves. Students must be encouraged to work creatively with new media, and while we might not know what to expect, it is likely that their work is not going to look anything like traditional academic work. Some of the arenas HUMlab will be working with in the immediate future include a space where drama history students will build theatres (from different periods), and a platform for medical index patients within a medical program. There are also well-progressed plans for a project where language students, engineers, media students and computer science students will look at Cyberspace from a multi-disciplinary and multi-cultural perspective.

Virtual worlds such as EVE provide us with places for learning, collaboration and representation. In describing the VW Project, we have focused on some constructive, integrative and creative aspects of such worlds. The project will continue next year and planned developments include an alumni program, a more radical 'plunge-in' approach, more focus on creative and artistic aspects and an extended student guide. One of the main differences between traditional assignments and the virtual 'products' is that there is no given endpoint in a virtual representation. The world remains, and the only real archive is the world itself. This is one of the reasons why all the students have expressed interest in the alumni program. It would be interesting to see how generations of student builders affect each other – both through what they have created and through personal meetings between students from different student groups. Thinking ahead, it would also be challenging to take the project a step further through creating another kind of dynamicity in the world. This could be done through introducing a more active environment through computer-generated cultural simulation with in-world robots and dynamic rules. And why not let the students play with creating a city full of moving, talking and thinking people, a wedding with a simulated step-mother or a contemporary monster (concrete or abstract) that affects and changes the whole world?

References

Anders, P. (1999). *Envisioning cyberspace: Designing 3D electronic spaces.* New York: McGraw-Hill.

Berger, A. A. (2002). *Videogames: A popular culture phenomenon.* New Jersey; Transactions Publishers.

Boisvert, R. D. (1998). *John Dewey: Rethinking our time.* Albany, New York: State University of New York Press.

Cherny, L. (1999). *Conversation and community: Chat in a virtual world.* Stanford: CSLI Publications.

Damer, B. (1998). *Avatars! Exploring and building virtual worlds on the internet.* Berkeley: Peachpit Press.

Danielsson, H. (2002). Att lära med media: Om det språkliga skapandets villkor i skolan med fokus på video[English title: Learning with media: Conditions for media language creation at school with a focus on video. Ph.D. thesis with summary in English]. Stockholm: Stockholm University.

Davis, E. (1998). *Techgnosis: Myth, magic + mysticism in the age of information.* London: Serpent's Tail.

Debski, R. (1997). Support of creativity and collaboration in the language classroom: A new role for technology. In R. Debski, J. Gassin & M. Smith (Eds.), *Language Learning through Social Computing.* Occasional Papers 16. Melbourne: Applied Linguistics Association of Australia & The Horwood Language Centre.

DeVarco, B. (2002). Presentation in *Active Worlds* on April 25, 2002.

Fosnot, C. T. (1996). Preface. In C. T. Fosnot (Ed.), *Constructivism: Theory, perspectives, and practice,* ix-xi. New York: Teachers College Press.

Gage, N. L. & Berliner, D. C. (1998). *Educational psychology.* Boston: Houghton Mifflin.

Gagnon, G.W. & Collay, M. (1996). *Constructivist learning design.* Unpublished paper. http://www.prainbow.com/cld/cldp.html.

Hayles, K. (1999). *How we became posthuman: Virtual bodies in cybernetics, literature and informatics.* Chicago: University of Chicago Press.

Lakoff, G. & Johnson, M. (1999). *Philosophy in the flesh: The embodied mind and its challenge to Western thought.* New York: Basic Books.

Lévi-Strauss, C. (1966). *The savage mind.* Chicago: University of Chicago Press.

Liggett, J. (1974). *The human face.* London: Constable.

Lunenfeld, P. (Ed.) (2000). *The digital dialectic: New essays on new media.* Cambridge, Massachusetts: MIT Press.

Markham, A. N. (1998). *Life online: Researching real experience in virtual space.* Walnut Creek: Rowman and Littlefield.

Massaro, D. W. (1998). *Perceiving talking faces: From speech perception to a behavioural principle*. Cambridge, Massachusetts: MIT Press.

McCloud, S. (1993). *Understanding comics: The invisible art*. New York: HarperCollins.

Rheingold, H. (1991). *Virtual Reality*. London: Mandarin.

Rheingold, H. (2001). Mobile virtual communities. *TheFeature*. http:// www.thefeature.com/.

Skinner, B.F. (1968). *The technology of teaching*. New York: Appleton-Century-Crofts.

Snyder, I. (2002). Silicon literacies. In I. Snyder (Ed.), *Silicon literacies: Communication, innovation and education in the electronic age*, 3-12. London: Routledge.

Svensson, P. & Ågren, P. O. (1999). Automater, bricoleurer och virtuella bröllop: tre paradigm för informationsteknik i undervisning [In Swedish. Translated title: Automats, bricoleurs and virtual weddings: Three paradigms for information technology in learning]. *HumanIT*, 3, 167-183.

Taylor, T. L. (2002). Living digitally: Embodiment in virtual worlds. In R. Shroeder (ed.), *The social life of avatars: Presence and interaction in shared virtual environments*, 40-52. London: Springer.

Turkle, S. (1995). *Life on the screen: Identity in the age of the internet*. New York: Simon and Schuster.

Websites

Achill Project – http://www.rheinahrcampus.de/international/projects/achill2000/
Adobe Atmosphere – http://www.adobe.com/products/atmosphere/
Active Worlds – http://www.activeworlds.com/
Traveler – http://www.digitalspace.com/Traveler/
Virtual Wedding Project – http://www.eng.umu.se/vw/

All websites cited in this chapter were verified on 28.08.2002.

Acknowledgements

Pat Shrimpton at the Department of Modern Languages at Umeå University has been involved in all the projects referred to here, and has also commented on a draft version of the chapter. Glen Ropella and Lesley Shield also contributed valuable comments. Per-Olof Ågren at the Department of Informatics at Umeå University helped contextualise the VW Project.

Appendix 1. Hypertext work on language attitudes (browser window).

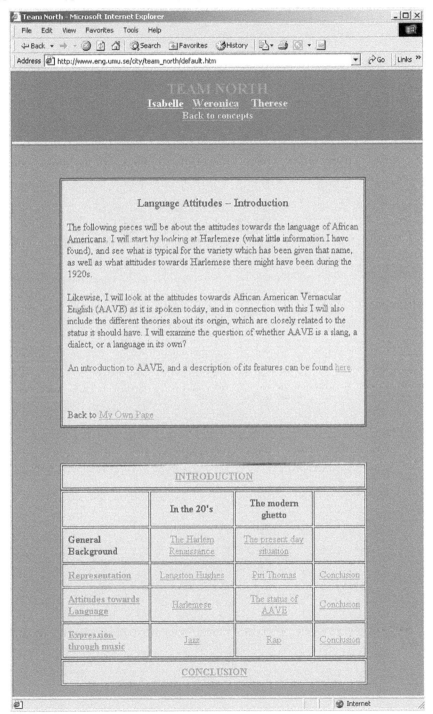

PEDAGOGY

8

Pedagogy on the line: identifying and closing the missing links

Uschi Felix, Monash University, Australia

Introduction

First the good news: in the years ahead, the declining cost of computation will make digital technologies, accessible to nearly everyone in all parts of the world, from inner-city neighborhoods in the United States to rural villages in developing nations. These new technologies have the potential to fundamentally transform how and what people learn throughout their lives. Just as advances in biotechnologies made possible the 'green revolution' in agriculture, new digital technologies make possible a 'learning revolution' in education.

Now, the bad news: while the new digital technologies make a learning revolution possible, they certainly do not guarantee it. Early results are not encouraging. In most places where new technologies are being used in education today, the technologies are used simply to reinforce outmoded approaches to learning. Even as scientific and technological advances are transforming agriculture, medicine, and industry, ideas about and approaches to teaching and learning remain largely unchanged. (Resnick 2002:32)

This is a commonly held view shared by experienced enthusiasts and informed critics of online learning. What is more, taken at face value, it is true. However, observations of this kind usually reflect one or two poor practices. The first is the transfer from a classroom setting to an online environment of learning materials developed for the still dominant knowledge transmission approach; the second, even more disappointing, is the replacement of pedagogically sound teaching with inferior online materials. In the latter context, we often hear the phrase 'we must radically change our teaching approach'. But is this in fact true? Are there significant differences between sound online and face-to-face pedagogy? Are there new proven online learning theories we can draw on? Of course, there is the relatively new field of instructional design to inform our decisions (Briggs 1977, Johnson & Foa 1989 and Gagné, Briggs & Wager 1992) but, just as in classroom settings, a large proportion of current materials are designed by the course providers and Berge's (1999:1) point is very important here:

How instruction is designed is based largely on the designer's interpretation of the world, filtered through his or her instructional philosophy. It is the instructional design, not the delivery system, that frequently sets the limits on the quality of instruction. The balance of the mix of various interpersonal interactions is a result of what the designer values.

The practice of experienced distance education institutions leaving course design in the hands of professional teams, including experts in content, pedagogy, instructional design and graphic design, makes a great deal of sense. But even then different teams will produce radically different courses depending on which educational theory they favour. Traditional computer-based learning has been designed with reference to theories developed by Ausubel (1963), Bandura (1986) and Gagné (1985) and much of today's pedagogy still adheres to Gagné's five categories of learning – intellectual skills, cognitive strategies, verbal information, motor skills, and attitudes (Doiron 2001). While there are notable exceptions, the majority of current online materials are based on this largely behaviourist approach and a view of learning as predominantly concerned with information processing.

Gagné's events of instruction include gaining attention, informing learners of the objective, stimulating recall of prior learning, presenting the new material (stimulus), providing learning guidance, eliciting performance, providing feedback, assessing performance, and enhancing retention and transfer.
Doiron (2001:2)

During the last decades there has been a move away from theories favouring behaviorist, individual cognitive approaches to those concerned with sociocultural constructivist ones, engaging students in problem solving, situated learning and co-operative activities. Interestingly this move does not reflect radically new philosophies in teaching but sees us 'marching backwards into the future', to borrow a phrase from Paulsen (1995), drawing inspiration from theorists such as Vygotsky (1978), Dewey (1963) and Leont'ev (1978) who emphasise interpersonal, experiential, process-oriented learning. It is important, though, to point out that the plethora of learning theories is so varied, and in some cases overlapping, that it would be naïve to suggest a black and white divide into strictly opposing schools of thought (see Duffy & Cunningham 1996 and Goodyear 2002). Perhaps the most significant, yet still fairly simplistic difference between exponents of Gagné's approach and those of Vygotsky's is that the former focus on the individual in the group, believing that cognition occurs in the head of the individual, whereas the latter emphasise the socially and culturally situated context of cognition, in which knowledge is constructed in shared endeavours (Duffy & Cunningham 1996). Generally, though, the fundamental questions to ask must surely be 'what are the principles of good practice in teaching?' and 'can all of them be realised online?'

Teaching must be grounded to basic principles of good practice in [...] education. These principles include encouraging contacts between students and faculty, developing reciprocity and co-operation among students,

encouraging active learning, giving prompt feedback, emphasising time on
task, communicating high expectations, and respecting diverse talents and
ways of learning. (Chickering & Gamson, 1987:3)

While these principles were realisable easily enough in face-to-face settings, honouring them in traditional distance education posed great challenges, especially for dedicated language teachers. The new technologies, however, provide the capabilities for doing all of this successfully, and in some instances offer the potential to improve even on best practice classroom teaching (Felix 2002a). The presence of authentic information gaps and the acute need for meaningful communication in distance education naturally shape the teaching approach towards a communicative model, and excellent teachers have always supplemented a largely text-based environment with personal communications conducted by telephone or through audio and video conferences. The advent of email and increasingly sophisticated interactive tools, including voiced chats and virtual worlds, have given teachers a much wider scope in which to engage students in best practice language learning, including facilitating communication with native speakers, creating real-life tasks carried out co-operatively in authentic settings, and addressing the large differences in students' backgrounds, interests, needs, learning strategies and abilities that are to be found in today's multicultural classrooms. The new technologies also offer more sophisticated and creative opportunities for the provision of timely feedback, although artificial intelligence capabilities have not advanced significantly enough to enable us to claim that automated versions alone would be sufficient in a course offered fully online.

Other important elements in good teaching, unfailingly cited by students of all ages as the characteristics of the teacher who had the greatest influence on them, are enthusiasm for the subject and humour. Sustaining these convincingly online is perhaps the greatest challenge but certainly not impossible. This author was greatly amused by a physics course entitled *Experimentil Erors* (Fallows & Ahmet 1999b) which clearly sets expectations way beyond the experience of the ordinary classroom, even though according to the tutor some students complained about poor spelling. Surely good teaching is good teaching in any setting, and a good teacher using a simple tool is likely more effective than a poor teacher using a highly sophisticated one, and there is no reason why online learning should not be as enjoyable as classroom equivalents (Hiss 2000).

Current stand-alone online language learning programs have improved tremendously over the last two years, especially through the recognition of the value of constructivist approaches in this environment, and increased interactivity in general. Naturally, teaching fully online adds a dimension which leaves no room for compromises in terms of infrastructure, resources, expertise, support and prerequisites discussed in detail in Felix (2002a). This is not the territory for the one-person enthusiast, although there are excellent examples of teachers who have dedicated their entire professional life to this challenge (see *Cyberitalian*). Large commercial enterprises such as *Global English* have invested heavily in interactivity online, always at the cutting edge of technology (including voice-recognition software) and employing 24-hour tutor support through a chat site. Generally, though, three important elements are still handled poorly in such offerings. These are *(1) providing personalised and meaningful feedback, (2) creating a sense of community and belonging*, and *(3) catering for the development of oral language skills*. This chapter looks at the fundamental pedagogical issues in the context of these important elements, and discusses innovative approaches for addressing the problems. It discusses innovative types of feedback structures and examines ways of personalising and humanising online learning in a systematic and holistic manner that will permeate the total student experience.

Feedback

The importance of feedback emerged very strongly in a series of studies carried out by the author on student perceptions of the web as a viable environment for language learning where it was expressed by students of all ages (Felix 2001b). This observation is supported by studies in other fields (Lyall & McNamara 2000, Sims 2000), as well as by earlier studies on distance education (Haughey 1990). It is also an element strongly emphasised in papers dealing with quality indicators for online learning (Illinois Report 1999, Ragan 1998, Kearsley 1998), and in discussions about creating interactivity and sustaining motivation (Goodfellow et al 1999, Labour 2001, Fallows & Ahmet 1999). Yet, despite compelling calls for meaningful, contextualised and, where possible, personalised feedback, a detailed survey of what is currently offered on the web in language learning (Felix 2001a) still shows a great preponderance of the drill-and-practice paradigm. There are several plausible reasons for this domination:

- The feedback freely available for us to look at online is not, and cannot be, a true indication of feedback in the courses concerned, since, in the vast majority of cases, it represents only part – and quite possibly a very small part – of what is being done. It is obvious that personal feedback provided through chat sites, bulletin boards, discussion groups, audiographics and email remains largely invisible to the casual visitor. Equally importantly, server-based applications in which the program knows how the user has performed previously and tailors the on-screen material accordingly, are by necessity restricted to enrolled participants.

- For reasons of ease and speed, teacher developers favour the use of simple templates and applets to produce student exercises, and this naturally limits what can be done. While many teachers are attempting to contextualise the work and are including the use of illustrative graphics, these exercises still largely represent drill-and-practice of various items with feedback restricted to right or wrong indications.

- Questions with only one correct answer – or at least a very limited number of alternatives – are not only the sort of activity that computers can mark most easily, but also constitute the sort of drudgery that teachers have long been tempted to transfer from humans to machines. A powerful drive here, therefore, is the desire to generate significant savings of time that would otherwise be put into marking.

The last two points clearly reflect an attitude that is still common: we ask what the computer can do for us, rather than what we can do with the technology (Felix 2002a). The result is that activities are accompanied by mechanical feedback on rote learning of facts, vocabulary and grammar. While this process has some merit – students like the instant feedback that computers can provide, and some of them will spend many hours at the screen in pursuit of the perfect score – it is far removed from providing the full repertoire of human feedback:

> *The key action for the educator is the use of positive and encouraging*
> *feedback. The nature of this feedback will vary with circumstance and will*
> *range from non-verbal communication (such as smiles and nods of the head)*
> *through oral praise (using encouraging words such as 'good' or 'excellent' as*
> *appropriate) to formal written comments on written work. Praise, as a tool to*
> *inspire students, is particularly effective if directly linked to the student's*
> *achievement of a specified learning outcome. For the less motivated student,*
> *there is also benefit in focusing praise on the effort put in since this will*
> *reinforce the link between the work undertaken and the achievement of the*
> *desired outcome. The key requirement is to build the understanding that*
> *achievement is not merely a matter of luck or a preordained inevitability.*
> *(Fallows & Ahmet 1999a·3)*

While these recommendations were made in the context of face-to-face teaching, there is no reason why online experiences should be more impoverished in terms of positive reinforcement. Overall, the proposed action might seem a tall order in an electronic environment, but in some instances the technology offers an advantage over what can be done in a classroom setting. Computers can provide individualised feedback when it is difficult in the classroom to attend to all students equally and fairly at all times; automated feedback can be more frequent and directly linked to very small achievements; a computer will tirelessly continue to give anonymous feedback, independent of moods and personal relationships, and independent of time and place. Of course, the more sophisticated the approach we adopt, the more resources in infrastructure and time will be required – a result that will seem paradoxical to those who embarked on computerisation in the first place in order to save time. There is one powerful compensation, however: if we do the work well, it will last a while and can serve as a model for other similar content. Another compensation is the enriched and personalised nature of the online experience that can be offered in various creative ways. We will now look in a little more detail at which features might be instrumental in the quest for best pedagogical practice in automated online feedback.

Hints

As well as hints before the exercise is attempted, a hint facility can be provided if the first response was wrong and a further attempt is allowed. Hints are not at all widespread, but it should be obvious that feedback that consists solely of marking the error in red – still less blanking it out – does little to help the student learn. While students appreciate the immediacy of automated feedback, they often complain about its quality (Felix 2000, 2001b) and some have specifically deplored the absence of useful online hints in Lanny & Musumeci (2000). Current sites display various forms of hints, which are conceptually so different from each other that it hardly makes sense to describe them by the same name. Although there is considerable overlap, we have broadly divided them here into structural and personalised types.

Structural hints

At a purely mechanical level, feedback uses pattern matching (by words, or by individual letter) to generate a report on how much of the response was correct. Ideally, all the elements that are correct will be identified. In practice, this is easier said than done. The simplest approach is to scan the response letter by letter (or word by word) from the beginning and display any matches. Some programmers stop the matching when the first error is found. Since this leads to a very small amount of help when the mistake occurs early in the answer, it might be better if the scan could be undertaken from the end of the response as well as from the beginning. The sort of activity in which pattern matching makes most sense is where the response is constrained. Requiring a set of scrambled words to be arranged in a sentence (by, for example, drag and drop) lends itself well to this approach, since all chance of misspelling is automatically eliminated.

Personalised hints

These contain personalised messages such as "well done, Sandra" or "oops, have another go Paul". More helpfully, they can offer detailed comments on why a response was incorrect with a reference to where more information can be found. Comments that are largely positive, even when the response is incorrect, give a much better feel to a program than a large bank of variations on the theme of "Wrong! Try again" (see Robinson 1991, Brandl 1995). However, it must be kept in mind that some students may feel patronised by anything other than being told that the answer was incorrect.

The hint can take the form of an explanation of the point at issue, with the object being to help the student engage in metacognitive reflection on the problem and use that understanding to produce a correct repeated response. This sort of feedback is going to be possible or, at least, highly accurate only for foreseen errors, though collecting student responses should allow for a rich database of likely errors to be built up over time. Two excellent examples of this type of hint have been developed by Heift (2001, 2002) in the context of teaching grammar, and by Pujolà (2001) to help with reading and listening comprehension. Heift's feedback is generated by an Intelligent Language Tutoring System (ILTS). The parser-based system analyses student input and provides error-specific feedback, exclusively in the target language, and includes a facility that matches feedback messages to learner expertise and provides remedial exercises (Heift 2001:99). The underlying pedagogy of the program reflects Garrett's (1987) discussion of the use of Natural Language Processing (NLP) for providing sophisticated feedback in which the individual explanation of errors mirrors more closely what might take place in a face-to-face setting. An interesting aspect of Pujolà's program is that it also provides a delayed two-step option that allows users to reflect on the reason for an error before accessing the explanation. Both authors have run evaluations on their systems which showed that the majority of learners do in fact make use of these hint facilities. Furthermore, Heift cites a number of studies that have found positive effects of metalinguistic feedback over traditional forms (Nagata 1995, 1996, Nagata & Swisher 1995). She also points to research by Van der Linden (1993) which showed that lengthy feedback messages are not being attended to and that feedback dealing with multiple errors was found to be too complex. Virvou et al (2000:13), who developed a similar program to Heift's for teaching the passive voice of English grammar, found that the 'Passive Voice Tutor was successful in achieving a high degree of compatibility with the human experts' opinion'.

Using a similar Intelligent CALL (ICALL) system, Chen & Tokuda (2002), Chen et al (2002) and Tokuda & Chen (2001) have developed a sophisticated program for online translation training based on template pattern matching. The templates use words or phrases as a minimal unit, with the databases selected by experienced language teachers in the light of responses collected from sample students. The program includes a heaviest common sequence algorithm for matches aimed at identifying, from among a large number of possible paths embedded within the template, the path with the greatest similarity to the learners' input translation. What the program delivers is error-contingent feedback for each student input.

While less sophisticated in terms of ICALL, an interesting approach to providing personalised feedback comes in *Arana's Spanish Language Exercises* which provide explanatory comments on correct answers as well as erroneous ones. Reinforcing success in this way seems a humane approach to the material. It is motivating, and it is an excellent way of increasing the illusion of personal contact. Not that comments do not sometimes irritate. Jovial responses to correct answers – whether the same one every time or one selected at random from a small list – can grate. Elaborate sound effects tend to be the most irritating versions in programs where students will hear the same sound repeatedly at every right or wrong turn. Neutral responses (Yes/Right or No/Sorry) look safer here, even if they are not 'interesting'. The problem in creating something more imaginative is striking the right tone for the audience. Testing various options with the target group is imperative here. Alternatively, graphics offer attractive options.

Graphics

Although research in this area is scarce, there is some evidence of positive effects of graphics used in teaching, especially in science subjects (Hedberg & Alexander 1994, Kaufmann et al 2000, Moore 1995 and Dalgarno & Hedberg 2001). Visual interface in general is seen as a fundamentally important element in online design (Boshier et al 1997, McGreal 1997, Jung 2001). Many different forms of integrating graphics into the learning resource are being used, ranging from simple stick figures to sophisticated simulations using virtual reality applications. To date there is no evidence that the latter produces better results than the former – investigations of this sort would contribute significantly to the field. Again, representation of graphics is broadly divided into structural and personalised types.

Structural graphics
These are generally simple pictures and drawings to illustrate a particular structural point or reinforce what is being introduced in textual and/or aural forms (Figure 1).

Multisensory input has long been held to benefit learning for two reasons. On the one hand different areas of the brain are stimulated simultaneously, on the other different learning styles (visual, auditory, symbolic) can be catered for (Dede 1996).

Figure 1. Grammar exercises – Hildes Hexenwelt

There are very few studies that have investigated the role of graphics specifically in language learning. Hew & Ohki (2001) found that Animated Graphical Annotation – a set of animated graphics representing visually the accent of Japanese words – was effective in improving students' listening skills. In our own studies we found a significant relationship between students' rating highly the usefulness of graphics (in this case simple stick figures) and working longer hours online (Felix 2001b). Amusingly, a participant at one of the author's presentations raised the question whether this could not simply have been the result of long download times! While this was not the case here, it is an important consideration in designing graphics. At no time should a desire for sophistication and look outweigh the pedagogical value of the item itself, especially not when functionality might be compromised.

Tokuda & Chen (2001) developed a part of speech tag parser which provides a visualised parsed tree of acceptable accuracy. This operates by applying the parser to the closest model translation to obtain the best matched path to the student input translation and identifying all deviations from the correctly parsed trees as errors. The parser not only provides a visualized tree demonstrating where students' errors are located but also acts as the base for setting up an efficient learner model showing how these errors can best be repaired (Chen et al 2002).

Personalised graphics

Accompanying characters acting as a tutor-companion feedback device have the potential to change the entire dynamics and climate of an online course. Pinocchio in Cyberitalian, Lina and Leo in the popular Goethe German course and the witch Hexe Hilde (Figure 2) in Hildes Hexenwelt add a personal dimension to interacting with a program that can add interest and curiosity and communicate humour and enthusiasm. It also helps in the creation of a larger and more varied repertoire of automated responses. Hexe Hilde who is young and quirky has her own special vocabulary which represents how German teenagers communicate these days. She can also swing a magic wand in approval or sweep away an incorrect response with her broom, appear as a contestant in a game show, and be the one to whom assignments are being sent. A word of caution: overly enthusiastic, gratuitous use of these, especially in animated versions, can become as irritating as sound effects; we only need to remember Microsoft's experience with Clippy, the animated Help sprite in Office 2000, which became so hated that it was abolished from Office XP.

Figure 2. Hexe Hilde – cybertutor extraordinaire

Scoring

In terms of student interest, there is good reason to provide a total score at the end – and quite possibly a running score as well when the questions are marked one by one. French@Austin has an attractive way of reporting the final results (correct, incorrect, not attempted) in a bar graph. Other sites store results and tap into student competitive spirit by printing out (say) the top 20 scores that have been achieved so far, or in the past day, week or month. It might seem that names will need to be collected to make the league tables most meaningful, but many students find this threatening and will prefer the safety of anonymity or pseudonymity and forego the public triumph.

Games

Other sites turn scoring into games. One example follows the model of Who wants to be a millionaire? with each successive correct answer visibly bringing the student closer to the goal. For true authenticity, a wrong answer should terminate the exercise, or, at least, in Weakest Link style, return the score to zero, but perhaps this would be too demotivating. Another uses a matrix of tiles to construct another quiz format, where the rows might cover a stated domain (like geography) with each tile in that row containing a question. This exercise can be played by one, but it can also be set up as a competition between two players who take turns at picking questions. These TV quiz formats are visual and familiar to many, so they make sense as the basis for games, if designers want to include such activities in their courses (Figure 3).

Figure 3. Wheel of fortune – Hildes Hexenwelt

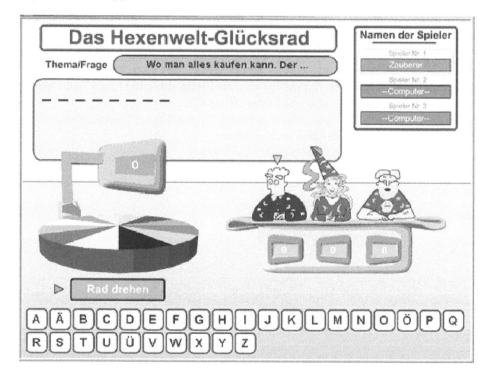

Community

One of the greatest concerns in distance education is low retention rates, with historical drop-out figures ranging from 30 to 50 percent (Hill 2001). Since distance education usually attracts non-traditional students (older students, students with spouses or with children at home and students working longer hours), a multitude of factors contribute to non-completion of courses. However, the important factors in whether or not students persevere with their studies appear most frequently to be the level and quality of interaction and support (Moore & Kearsley 1996, Hara & Kling 1999). While the new networked technologies have the potential to engage distance students in high levels of interaction and support, it is a challenging task to build and sustain online communities in an evolving and dynamic environment (Costigan 1999), which needs careful planning and facilitation during design, development and implementation. It is important to make explicit to learners the importance of community and how it might contribute to the learning (Hill 2001), and to involve learners in creating and sustaining community. As with all educational settings, heterogeneity of groups needs to be taken into account, and the fact that online learning potentially increases individual differences in terms of learning styles and strategies does not simplify the task.

However, there are two major ways in which a sense of community and belonging might be generated, taking into account different interests, abilities, backgrounds, learning styles and strategies. On the one hand, students might be given the opportunity to contribute individually to both the learning resource and the social environment in which the course is situated. This could consist of creating simple exercises and activities using templates such as *Hot Potatoes* or *Quia*, or generating ideas for task-based web quests or mystery games which would form part of the learning materials used by the group as a whole. Students could also post personal biographies, including photos, graphics, drawings, and so on to introduce themselves to the group. In our own German course we have given students the option of adopting fictional characters in the style of the witches' world of *Hexe Hilde*. On the other hand, the literature shows a growing body of successful examples in which a sense of belonging to a community might be created by students co-operating and collaborating in experiential and constructivist projects, using a variety of communication facilities ranging from mystery quests using simple email to building communities in virtual worlds (Warschauer 1995, 1996, Debski et al 1997, Warschauer & Kern 2000, Felix 2002a). Typically students would be involved in communicating with native speakers in the target language country, collaborating on negotiated projects, carrying out research tasks on authentic websites, producing websites on topics of interest, participating in chats and MOOs, or engaging in interactive simulation exercises (i.e. taking a virtual trip to a country with various choices along the way relating to budget, destinations, purchases, sending a greeting card etc.). Since chapters six and seven discuss some of these in detail, we will not describe more projects but concentrate here on the potential problems associated with co-operative ventures and on ways of overcoming them.

In his excellent introduction to Dewey's work, Campbell (1995) highlights the evaluative power of intelligence as a possession of the group rather than of individuals. This is a powerful incentive to include collaborative work in any curriculum. However, four important elements need to be considered in the quest for success. The first is dealing with the consequences of what Kollock & Smith (1996) term *social dilemma*, the second is striking a balance between teacher and learner control, the third is designing assessment procedures which are both authentic and fair to all, and the last but not least is managing time.

Social dilemma

> *At the root of the problem of co-operation is the fact that there is often a*
> *tension between individual and collective rationality. That is to say that in*
> *many situations, behaviour that is reasonable and justifiable for the*
> *individual leads to a poorer outcome for all (Kollock & Smith 1996, cited in*
> *Goodyear 2002:69)*

This is reminiscent of Bandura's view that 'inveterate self-doubters are not easily forged into a collectively efficacious force' and that 'perceived collective efficacy will influence what people choose to do as a group, how much effort they put into it, and their staying power when group efforts fail to produce results' (Bandura 1986:449). These sorts of problems are often cited in discussions of collaborative ventures (Felix 1999, Levy 1997), and overcoming them requires careful attention to group dynamics and unobtrusive monitoring of group activities on the part of the teacher. Goodyear (2002:70) further cites Kollock who draws on research which has studied whole communities acting together. Communities that have been successful in managing collective resources and social dilemmas have had the following features:
1. A clearly defined group boundary, to help identify who is making use of group resources.
2. Local customization of norms and rules about the use of group resources (enabling people in the group who were knowledgeable and committed to have a say in shaping the group's rules).
3. A system for monitoring and sanctioning members' behaviour (without recourse to an external authority).
4. Access to low cost conflict-resolution mechanisms.

Learner versus instructor control

These points highlight the importance of creating a balance between learner control and instructor control in online learning. The concept of Moore's (1983) *transactional distance* is relevant here. This key concept in distance education theory is defined as a function of two variables, *dialogue* and *structure* in which 'distance in education is not determined by geographic proximity, but rather by the level and rate of dialogue and structure' (Saba & Shearer 1994). In an interesting empirical study, using a form of desk-top video-conferencing, the latter authors set out to verify this concept and found the following:

> *An increase in the level of learner control increased the rate of dialogue,*
> *which in turn decreased the level of transactional distance; an increase in the*
> *level of instructor control increased the rate of structure, which in turn*
> *increased the level of transactional distance. (Saba & Shearer 1994:54)*

Naturally, there cannot be any hard and fast rules that will apply to all students equally. While some students prefer working unguided and in groups, others need more structure and frequent reinforcement from the teacher. A neat idea for reducing uncertainty about individual involvement in project work has been suggested by Murphy, Mahoney & Harvell (2000) who designed explicit contracts for instructor and student responsibilities.

The variable of scepticism has also been found to be influencing student participation online, for example if students strongly believe that they will not forge any relationships online, they will not (Utz 2000). Another important element in group dynamics is students' self-concept about their abilities: 'Among the types of thoughts that affect action, none is more central or pervasive than people's judgements of their capabilities to deal effectively with different realities' (Bandura 1986:21). The teacher's role in communicating positive expectations is no less important online than it is in a classroom. This can often be handled in private communications between teacher and student (see next section), away from the core business of the project work. It is also an important part of dealing with the assessment of co-operative projects.

Authentic assessment

If we believe in testing what we teach it becomes obvious that current online automated testing devices, designed to teach objective knowledge, are largely unsuited for the purpose of testing what the students have learnt in the processes that accompany constructivist approaches. While there is a place for such forms of testing in formative assessment along the way, fostering a reflective process on the more formal structures of the language that have been acquired, the majority of assessment procedures need to focus on both the process and the outcome.

The most distinct feature of authentic testing is that the emphasis is not simply on testing knowledge summatively but that the assessment process itself is seen as a continuation of the learning process and naturally accompanies the learning program in a variety of ways. This is a great challenge. While it is fairly easy to set up creative and meaningful real-life learning situations online, testing all of it online (and automatically) is pretty much impossible in the absence of sophisticated intelligent tutor systems. The following are typically used in authentic assessment:

Personal Journals (on/off line – formative)
  • for process oriented reflective purposes and self-assessment

Peer assessment (online – formative+summative – may include expert)
  • in discussion groups, BBS (with sound), MOOs, Chats,
  • judging of short answer questions
  • judging of contributions to larger topics, interactive stories

Portfolios (on/off line – formative+summative – marked by tutor)
  • dynamically evolving measure
  • includes outcomes of various tasks, processes, interactions negotiated between tutor and students

Products (online – summative – marked by tutor/peer/expert)
  • tangible outcomes of co-operative or sole project
  • websites produced on specific topics

Automated feedback (online – formative)
  • sophisticated versions that give hints, refer to relevant help, foster learning

Time management

One of the most common complaints by even the most enthusiastic advocates of the types of learning activities discussed in this and the following section is that it is very time-consuming (see Felix 2002b). It is also clear that authentic assessment is not a time-saving device and that procedures need to be discussed and agreed to by the group as a whole. Peer assessment can be especially problematic if students are not fully comfortable with the process (Mason 1993). However, there are a few time management devices that practitioners have found useful. First, students need to be given clear guidelines as to what is expected from them and how much tutor access is reasonable. Secondly, help from volunteer native speakers in the role of mentors, monitors, assessors might be enlisted. Thirdly, projects must not be overly ambitious but reflect the collective skills of individual groups; it is important that each student has the opportunity to contribute equally to a project in terms of effort, albeit in different intellectual and practical ways. A mark can then be given to the project as a whole with points added or deducted according to individual contributions as agreed to by the group. While students sometimes resist this approach initially, in this author's experience they usually cope well after a while, and it is often surprising how fairly students assess each other's efforts when a sense of community has been established. Fourthly, summative assessment of structures can be handed over to the computer, consisting of a random bank of exercises that students have worked with along the way. Finally, the quest for excellent pedagogy should not lead to overloading students with work. There in no reason why we should demand more of students in online environments than in the classroom. Finding a balance between recognising individual students' needs and providing support to the group as a whole will be a continuing challenge.

Speaking online

The difficulty of catering for the development of oral production skills has long been deplored in distance language teaching (Abrioux 1991, Williams & Sharma 1988), and not much has changed since these observations were made. Oral activities are still conspicuously absent from online offerings, and students' complaints about this usually head the list of disadvantages associated with web-based language learning (Felix 2001b). However, during the last few years practitioners have begun to incorporate sophisticated applications in the form of audiographics (see chapter nine), voice recognition software (*Global English*), voiced bulletin boards (*Wimba*) and voiced chats (*Traveler*). We will here concentrate on the last two since they have most to offer in the context of our discussion of creating meaningful constructivist activities and contributing to a sense of community and belonging. But before taking a look at the potential of these applications, let us briefly discuss a fundamental problem in the development of oral language skills in traditional face-to-face settings, namely *language anxiety*.

Experienced teachers know that it can be difficult to motivate some students to speak in class. Speaking often remains the least accomplished of the four language skills, especially in older students and those who have never spent any time in the target language country. One of the reasons for this is that the development of oral skills carries a high level of ego involvement, especially when students feel exposed in front of an entire class, and that 'in conditions of high ego involvement, anxiety has typically been found to interfere with performance' (Sinclair 1971:96). While findings of the effects of anxiety on language learning in particular are equivocal, though leaning towards a negative effect, the presence of anxiety in speaking a foreign language in public is well documented (Horwitz & Young 1991). The phenomenon of 'linguistic avoidance' is especially often observed during oral exams when anxious students avoid certain linguistic structures and topics (Kleinmann 1977, Steinberg & Horwitz 1986).

The question here is whether anonymous environments might improve this situation by making students feel less anxious. Since there is no specific research on this topic as yet, it is useful to look at other environments where synchronous and asynchronous communications have been used in educational settings, albeit in written form.

> *...some types of trainees who do not do well in spontaneous spoken interaction (e.g. students who are shy, reflective and more comfortable with emotional distance) find that asynchronous, text-based communication better fits their learning style. For this person informal written communication via computer conferencing is often more authentic than face-to-face verbal exchange...(Dede 1996:17)*

A striking example of the benefits of anonymity on an otherwise reticent and anxious student was reported by Freeman & Capper (1999). In this postgraduate business course, in which students chose an alias to conduct class business exclusively online, a very shy female student of Muslim background transformed herself into a powerful contributor to the group under the pseudonym of the Australian Prime Minister. Although an extreme example, it suggests that the simultaneous freedom of making mistakes in a safe environment and the pseudo authenticity of the task may well have a salutary effect, a speculation supported by Roberts et al (1996). It has been noted, though, that learners participating in these environments may also display negative disinhibitions in the form of 'flaming', hurling insults they would never use face-to-face (Sproull & Kiesler 1991, quoted in Dede 1996:27). This highlights the need for clear guidelines about possible consequences of inappropriate communications, outlined in the previous section.

Benefits specifically related to language learning of computer assisted classroom discussions (CACD), as repeatedly singled out in the literature, are listed in an excellent article by Ortega (1997:83):

- Learners are able to contribute as much as they want at their own pace and leisure; consequently, they tend to perceive CACD as less threatening and inhibiting than oral interactions and produce a high amount of writing, with all students participating to a high degree and all producing several turns/ messages per session.

- Because of the interactive nature of the writing, learners are expected to engage in a variety of interactive moves on the computer and to take control of managing the discussion.
- Learners make use of the available opportunity to take time to plan their messages and edit them. In this way they engage in productive L2 strategies and processes.
- Learners have exposure to a substantial amount of comprehensible input which is produced by peers of a similar level and shared background.
- Learners get a considerable amount of reading practice in addition to writing practice. Because of at least two tasks (writing and reading) competing for the learner's investment, reading skills practised may tend to be holistic (reading for the gist) and meaning-driven. In addition, learners are expected to be motivated to read because of an authentic sense of interactive audience provided by CACD.

In this context can we speculate that the use of *Wimba*, a voiced threaded bulletin board, in which students can be engaged in listening, speaking, reading and writing, might have similar or even greater benefits? In the absence of rigorous research we cannot make any claims here, but what is clearly the case is that the conditions for reading and writing hold. Whether the addition of speaking and listening interferes or enhances the process remains to be tested. In our own course we have incorporated the *Hexe Hilde* character into *Wimba* to retain consistency of *Hilde* as companion.

Figure 4. Wimba voice board – Hildes Hexenwelt

Wimba provides excellent potential for extending the sorts of activities referred to in the previous section to include speaking. Rather than using this very user-friendly application for traditional pronunciation practice, it could be used for creative information gap exercises, mystery games, the creation of interactive stories, reorganising mismatched oral and written cues, and many more, depending on the imagination of the teacher and participating students. Here, too, involving students in the setting up of tasks and exercises may be of great benefit. An excellent aspect of Wimba is that it also contains a facility for sending voiced emails, which offers opportunities for private communications between teacher and students and between the students themselves. Students, therefore, can be given various options for oral communication: publicly or anonymously via the bulletin board, or via email, or through any chosen combination. The fact that voices might be recognised may be a slight obstacle to real anonymity, but it is not too hard to disguise a voice somewhat, especially if one's chosen identity is *bear* or *magician*, and the possible benefits of simply adopting an alias has long been observed in experimental language classes in face-to-face settings (Felix 1989).

The greatest strength of *Wimba* is its simplicity of use. All that students need is a sound card in their computer, a microphone and minimal instructions on how to use the facility which can be run from the *Wimba* server or installed on a dedicated local server. Sound quality varies according to the sophistication of both, but in general a fairly recent PC with a relatively good microphone seems to suffice. It is important to point out, though, that sound quality does not match the quality of RealAudio; it uses the same compression format as mobile phones. Still, the flexibility of the resource for creative activities more than compensates for the difference.

Much higher quality of sound is offered in the synchronous voiced chat *Traveler* (Figure 5). This very ambitious site offers state-of-the-art avatars through which to communicate either at a set time with other class members and planted native speakers or with anyone who happens to be in the chat at the time. For the teaching of ESL the latter may be motivating and useful, albeit only at a fairly high level of proficiency, since common users tend to be native speakers of English. *Traveler* is not as user-friendly as *Wimba*. To set it up may require liaison with network personnel where firewalls exist.

Figure 5. Traveler avatars

These are very challenging environments, of course. While they offer learners the possibility of anonymity and the opportunity to make mistakes in an unthreatening and entertaining environment, they pose several serious problems that need to be addressed by teachers before embarking on activities. First of all, there is the claim that synchronous communications can restrict students (Berge 1999). This is especially true with learners of another language at lower levels of proficiency. While the environment may well be anonymous, it does not allow for the luxury of careful composing, reflection and multiple re-recording of the asynchronous *Wimba*. Rather than throwing students into such an environment at the deep end, a clear need for its use has to be established. The advantage that *Traveler* offers over *Wimba* is that it provides authenticity both of task and setting. While *Wimba* lends itself well to structured *learning* activities, *Traveler* offers opportunities for risk-taking and unplanned communication with native speakers under real-life speaking conditions, dealing with authentic information gaps. The price to pay for this authenticity, however, can sometimes be inappropriate and unwanted communications generated by dubious anonymous characters, seriously compromising Kollock's proviso of a clearly defined group boundary in successful management of collective resources. While adults may cope with such intrusions easily enough, in school environments negotiation of a private channel may be advisable. This will reduce authenticity but allow for well planned, small group interaction around a set task, say a debate or a short play, to which native speakers might be invited as contributors, monitors or arbitrators. An idea for using both applications for different purposes would be to produce an interactive story or play on *Wimba* and then act it out in *Traveler*. A wonderful feature of both environments is that they allow students to choose between anonymity and public exposure, and to move freely between real and imaginary worlds, which not only caters for different learning styles and preferences but also offers real opportunities for reducing language anxiety.

> *...participants in synthetic environments often feel as if the machine-based agents they encounter are real human beings, an illustration of the general principle that users tend to anthropomorphize information technologies (Weitzenbaum 1976). As a complement to responding to knowbots as if they were human, participants in a virtual world interacting via avatars tend to treat each other as imaginary beings. (Dede 1996:26)*

A final observation: In all these endeavours it is important that the technology does not dominate the learning experience but remains in the background in the shape of one of many tools at the disposal of both teachers and students, used for the unique potential it offers in different settings and in catering for different learning needs.

Conclusion

It can be said with some confidence that identifying a single best-practice pedagogical approach for online learning is impossible. After all, important elements in good teaching are the personality of the teacher, the needs of the particular student cohort, and not least the characteristics of the setting in which teaching and learning take place. What is significantly different in an online environment, when compared to a classroom setting, is the fact that we are dealing simultaneously with radically different approaches to providing our students with materials, feedback and opportunities for interaction. On the one hand, we have the ability to expose learners to reasonably sophisticated automated activities that will engage them in autonomous, predominantly cognitive and metacognitive processes, informed by theory drawing on the work of Gagné. On the other, we are in a position to exploit the unique opportunities of networked systems to engage students in authentic constructivist learning, in which students interact and collaborate in process-oriented real-life activities, informed by theorists such as Vygotsky and Dewey. Although quite different, the two schools of thought complement each other well in an online environment, especially if we take some care to humanise and personalise the former as much as possible within current technological limitations. In contrast to separating CALL activities and interpersonal activities, which is the predominant mode in face-to-face teaching, providing all activities through the same medium, and involving students in the creation of some of these, may well produce a more consistent climate of community and belonging than we find in traditional distance education and perhaps even some classrooms. To test this bold assertion will make a rich topic for further research.

References

Abrioux, D. (1991). Computer-Assisted Language Learning at a Distance: An International Survey. *American Journal of Distance Education,* 5 (1), 3-14.

Ausubel, D.P. (1963). *Psychology of Meaningful Verbal Learning.* New York: Grune & Stratton.

Bandura, A. (1986). Social foundations of thought and action: A social cognitive theory. Upper Saddle River, NJ: Prentice-Hall.

Berge, Z. (1999). Interaction in Post-Secondary Web-Based Learning. http://www.saskschools.ca/~parkland/interaction.htm

Boshier, R., Mohapi, M., Moulton, G., Qayyum, A., Sadownik, L. & Wilson, M. (1997). Best and worst dressed Web lessons: Strutting into the 21st century in comfort and style. *Distance Education,* 18 (1), 327-349.

Brandl, K. K. (1995). Strong and Weak Students' Preference for Error Feedback Options and Responses. *The Modern Language Journal,* 79 (I), 194-211.

Briggs, L.J. (1977). *Instructional Design Principles and applications.* Englewood Cliffs, M.J.: Educational Technology Publications.

Campbell, J. (1995). *Understanding John Dewey. Nature and co-operative intelligence.* Chicago: Open Court.

Chen,L. & Tokuda, N. (2002). Bug diagnosis by string matching: Application to ILTS for translation. In press *CALICO,* 20 (2).

Chen, L., Tokuda, N. & Xiao, D. (2002). A POST Parser-Based Learner Model for Template-Based ICALL for Japanese-English Writing Skills. In press, *Special Issue on ICALL for CALL.*

Chickering, A.W. & Gamson, Z. (1987). Seven principles for good practice in undergraduate education. *AAHE Bulletin,* 39 (7), 3-7.

Costigan, J.T. (1999). Forests, trees, and Internet research. In S. Jones (Ed.), *Doing Internet research: Critical issues and methods for examining the net,* xvii-xxiv. Thousand Oaks, CA: Sage.

Dalgarno, B. & Hedberg, J. (2001). 3D Learning Environments in Tertiary Education. Conference Proceedings, *ASCILITE '01: Meeting at the Crossroads*, Melbourne, Victoria, 33-36.

Debski, R., Gassin, J. & Smith, M. (Eds.) (1997). *Language Learning through Social Computing.* Occasional Papers 16. Melbourne: Applied Linguistics Association of Australia & The Horwood Language Centre.

Dede, C. (1996). The Evolution of Distance Education Emerging Technologies and Distributed Learning. *The American Journal of Distance Education,* 10 (2), 4-36.

Dewey, J. (1963). *Experience and Education.* New York: Collier Books.

Doiron, J.A. (2001). e-Education a 2001 Cyber-Space Odyssey? *CDTL Brief* 4 (3), 1.

Duffy, T.M. & Cunningham, D.J. (1996). Constructivism: Implications for the design and delivery of instruction. In D. Jonassen (Ed.), *Handbook of Instructional Technology.* New York: Simon & Schuster Macmillan.

Fallows, S. & Ahmet, K. (Eds) (1999). *Inspiring students: Case studies in motivating the learner.* London: Kogan Page.

Fallows, S. & Ahmet, K. (1999a). Inspiring students: An Introduction. In S. Fallows & K. Ahmet (Eds), *Inspiring students: Case studies in motivating the learner*, 1-5. London: Kogan Page.

Fallows, S. & Ahmet, K. (1999b). Editors' concluding comments. In S. Fallows & K. Ahmet (Eds), *Inspiring students: Case studies in motivating the learner*, 169-173. London: Kogan Page.

Felix, U. (1989). An Investigation of the Effects of Music, Relaxation and Suggestion in Second Language Acquisition in Schools. Unpublished PhD thesis, Flinders University, Adelaide, Australia

Felix, U. (1999). Web-Based Language Learning: A Window to the Authentic World. In R. Debski & M. Levy (Eds.), *WORLDCALL: Global Perspectives on Computer-Assisted Language Learning,* 85-98. Lisse: Swets & Zeitlinger.

Felix, U. (2000). The potential of CD-ROM technology for integrating language and literature: student perceptions. *German as a Foreign Language*, September. http://www.gfl-journal.de/previous/index.html

Felix, U. (2001a). *Beyond Babel: Language Learning Online.* Melbourne: Language Australia.

Felix, U. (2001b). The Web's potential for language learning: The student's perspective. *Recall,* 13 (1), 47-58.

Felix, U. (2002a). The Web as vehicle for constructivist approaches in language teaching. *Recall,* 14 (1), 2-15.

Felix, U. (2002b). Teaching languages online: Deconstructing the myths. Paper presented at the *Setting the agenda: Languages, Linguistics and Area Studies in Higher Education Conference, Manchester*. In press Conference Proceedings, London: CILT.

Freeman, M. & Capper, J. (1999). Exploiting the Web for education: An anonymous asynchronous role simulation. *Australian Journal of Educational Technology*, 15 (1), 95-116.

Gagné, R.M. (1985). *The Conditions of Learning and Theory of Instruction.* New York: Holt, Rinehart and Winston.

Gagné, R.M., Briggs, L.J. & Wager, W. (1992). *Principles of instructional design.* Fort Worth, TX: HBJ College Publishers.

Garrett, P. B. (1987). A Psycholinguistic Perspective on Grammar and CALL. In W.F. Smith (Ed.), *Modern Media in Foreign Language Education: Theory and Implementation,* 169-196. Lincolnwood: National Textbook.

Goodfellow, R., Manning, P. & Lamy, M-N. (1999). Building an online open and distance language learning environment. In R. Debski & M. Levy (Eds.), *WORLD CALL Global Perspectives on Computer-assisted language learning,* 267-285. Lisse: Swets & Zeitlinger.

Goodyear, P. (2002). Psychological foundations for networked learning. In C. Steeples & C. Jones (Eds.), *Networked Learning: Perspectives and Issues,* 49-75. London: Springer.

Hara N. & Kling, R. (1999). Students' frustrations with a Web-based Distance Education Course. *First Monday,* 4 (12), 1-33. http://firstmonday.org/issues/issue4_12/hara/index.html

Haughey, M. (1990). Distance Education in Schools. *The Canadian Administrator,* 29 (9).

Hedberg, J.G. & Alexander, S. (1994). Virtual Reality in Education: Defining researchable issues. *Educational Media International,* 31 (4), 214-220.

Hcift, T. (2001). Error-specific and individualised feedback in a Web-based language tutoring system: Do they read it? *ReCall,* 13 (1), 99-109.

Heift, T. (2002). Learner Control and Error Correction in ICALL: Browsers, Peekers and Adamants. *CALICO,* 19 (3), 295-313.

Hew, S. & Ohki, M. (2001). A study on the effectiveness and usefulness of animated graphical annotation in Japanese CALL. *ReCALL,* 13 (2), 245-260.

Hill, J.R. (2001). Building Community in Web-Based Learning Environments: Strategies and Techniques. Paper delivered at *AusWeb01,* Coffs Harbour, Australia.

Hiss, A. (2000). Talking the Talk: Humor and Other Forms of Online Communication. In K.W. White & B.H. Weight (Eds.), *The Online Teaching Guide: A Handbook of Attitudes, Strategies, and Techniques for the Virtual Classroom,* 24-36. Boston: Allyn and Bacon.

Horwitz, E. & Young, D.J. (Eds.) (1991). *Language Anxiety: From Theory and Research to Classroom Implications.* New Jersery: Prentice Hall.

Illinois Report (1999). *Teaching at an Internet Distance: the Pedagogy of Online Teaching and Learning.* The report of a 1998-1999 University of Illinois Faculty Seminar. http://www.vpaa.uillinois.edu/tid/report/toc.html

Johnson, K.A. & Foa, L. J. (1989). *Instructional Design: New Alternatives for Effective Education and Training.* New York: Macmillan.

Jung, I. (2001). Building a theoretical framework of Web-based instruction in the context of distance education. *British Journal of Educational Technology,* 32 (5), 525-534.

Kaufmann, H., Schmalstieg, D. & Wagner, M. (2000). Construct3D: A Virtual Reality Application for Mathematics and Geometry Education. *Education and Information Technologies,* 5(4), 263-276.

Kearsley, G. (1998). Online Education: New Paradigms for Learning and Teaching. http://horizon.unc.edu/TS/vision/1998-08.asp

Kleinmann, H.H. (1977). Avoidance behavior in adult second language acquisition. *Language Learning,* 27, 93-107.

Kollock, P. & Smith, M. (1996). Managing the virtual commons: Cooperation and conflict in computer communities. In Herring, S. (Ed.), *Computer mediated communication: Linguistic, social, and cross cultural perspectives,* 109-128. Amsterdam: John Benjamins.

Labour, M. (2001). Social constructivism and CALL: Evaluating some interactive features of network-based authoring tools. *ReCall,* 13 (1), 32-47.

Lanny, A. & Musumeci, D. (2000). Instructor attitudes within the SCALE Efficiency Projects. *Journal of Asynchronous Learning Network,* 3 (4). http://www.aln.org/alnWeb/journal/Vol4_issue3/fs/arvan/fs-arvan.htm

Leont'ev, A. N. (1978). *Activity, consciousness, and personality.* M.J. Hall, trans. Englewood Cliffs, NJ: Prentice Hall.

Levy, M. (1997). Project-based learning for language teachers: Reflecting on the process. In R. Debski, J. Gassin & M. Smith (Eds.), *Language learning through social computing,* 181-199. Occasional Papers 16. Melbourne: Applied Linguistics Association of Australia (ALAA) and the Horwood Language Centre.

Lyall, R. & McNamara, S. (2000). Learning tool or potplant stand? Students' opinions of learning from a CAL program in a distance education context. *Australian Journal of Educational Technology,* 16 (2), 126-146.

Mason, R. (1993). Designing collaborative work for online courses. In G. Davies & B. Samways (Eds.), *Teleteaching,* 569-578. Proceedings of the IFIP TC3 Conference, *TeleTeaching 93,* Trondheim, Norway.

McGreal, R. (1997). Information technology and telecommunications: A course on the WWW. *Journal of Distance Education,* 12 (1/2), 67-84.

Moore, M.G. (1983). The individual adult learner. In M. Tight (Ed.), *Adult Learning and Education,* 153-168. London: Croom Helm.

Moore, M.G. & Kearsley, G. (1996). *Distance education: A systems view.* New York: Wadsworth.

Moore, P. (1995). Learning and teaching in virtual worlds: Implications of virtual reality for education. *Australian Journal of Educational Technology,* 11 (2).

Murphy, K.L., Mahoney, S.E. & Harvell, T.J. (2000). Role of Contracts in Enhancing Community Building in Web Courses. *Educational Technology & Society,* 3 (3), 1-18.

Nagata, N. (1995). An Effective Application of Natural Language Processing in Second Language Instruction. *CALICO,* 13 (1), 47-67.

Nagata, N. (1996). Computer vs. Workbook Instruction in Second Language Acquisition. *CALICO Journal,* 14 (1), 53-75.

Nagata, N. & Swisher M. V. (1995). A Study of Consciousness-Raising by Computer: The Effect of Metalinguistic Feedback on Second Language Learning. *Foreign Language Annals,* 28 (3), 337-347.

Ortega, L. (1997). Processes and outcomes in networked classroom interaction: Defining the research agenda for L2 computer-assisted classroom discussion. *Language Learning & Technology,* 1 (1), 82-93. http://llt.msu.edu/vol1num1/ortega/default.html

Paulsen, M.F. (1995). The Online Report on Pedagogical Techniques of Computer-Mediated Communication. http://www.nettskolen.com/forskning/19/cmcped.html

Pujolà, J-T. (2001). Did CALL feedback feed back? Researching learners' use of feedback. *ReCall,* 13 (1), 79-99.

Ragan, L.C. (1998). Good Teaching is Good Teaching: An Emerging Set of Guiding Principles and Practices for the Design and Development of Distance Education. *DEOSNEWS,* 8 (12). http://www.ed.psu.edu/acsde/deos/deosnews/deosnews8_12.asp

Resnick, M., (2002). Rethinking learning in the Digital Age. In G.S. Kirkman (Ed.), *Global Information Technology Report 2001-2002: Readiness for the Networked World.* Oxford: Oxford University Press.

Roberts, L.D., Smith, L.M. & Pollock, C. (1996). Social Interaction in MOOs: Constraints and Opportunities of a Text-Based Virtual Environment for Interpersonal Communication. http://wwwmcc.murdoch.edu.au/ReadingRoom/VID/VIDpap.html

Robinson, G. L. (1991). Effective Feedback Strategies in CALL. Learning theory and Empirical Research. In Dunkel, P. (Ed.), *Computer-Assisted language and testing,* 155-167. New York: Newbury House.

Saba, F. & Shearer, R. (1994). Verifying key theoretical concepts in a dynamic model of distance education. *American Journal of Distance Education,* 8 (1), 36-56.

Sims, R. (2000). An interactive conundrum: Constructs of interactivity and learning theory. *Australian Journal of Educational Technology,* 16 (1), 45-57.

Sinclair, K. E. (1971). The influence of anxiety on several measures of performance. In Gaudry & Spielberger (Eds.), *Anxiety and Educational Achievement,* 95-106. Sydney: Wiley & Sons.

Sproull, S. & Kiesler, S. (1991). *Connections: New Ways of Working in the Networked World.* Cambridge, MA: MIT Press.

Steinberg, F.S. & Horwitz, E.K. (1986). The effect of induced anxiety on the denotative and interpretative content of second language speech. *TESOL Quarterly,* 20, 131-136.

Tokuda, N. & Chen, L. (2001). An Online Tutoring System for Language Translation, *IEEE Multimedia,* 8 (3), 46-55.

Utz, S. (2000). Social information processing in MUDs: The development of friendships in virtual worlds. *Journal of Online Behavior,* 1 (1). http://www.behavior.net/JOB/v1n1/utz.htm

Van der Linden, E. (1993). Does Feedback Enhance Computer-Assisted Language Learning? *Computers & Education,* 21 (1-2), 61-65.

Virvou, M., Maras, D. & Tsariga, V. (2000). Student Modelling in an Intelligent Tutoring System for the Passive Voice of English Language. *Educational Technology & Society,* 3 (4). http://ifets.massey.ac.nz/periodical/vol_4_2000/virvou.html

Vygotsky, L.S. (1978). *Mind in Society.* Cambridge, MA: Harvard University Press.

Warschauer, M. (1995). *Virtual Connections: Online Activities & Projects for Networking Language Learners.* Hawai'i: University of Hawai'i Press.

Warschauer, M. (1996). Comparing face-to-face and electronic discussion in the second language classroom. *CALICO Journal,* 13, 7-25.

Warschauer, M. & Kern, R. (Eds.) (2000). *Network-based Language Teaching: Concepts and Practice*. Cambridge: Cambridge University Press.

Weitzenbaum, J. (1976). *Computer Power and Human Reason*. San Francisco: W.H. Freeman.

Williams, S. & Sharma, P. (1988). Language acquisition by distance education: An Australian survey. *Distance Education, 9*, 127-146.

Websites

Cyberitalian – http://www.cyberitalian.com/
French@Austin – http://www.lamc.utexas.edu/fr/home.html
Global English – http://www.globalenglish.com/
Hilde's Hexenwelt – under construction
Hot Potatoes – http://web.uvic.ca/hrd/hotpot/
Lina und Leo – http://www.goethe.de/z/50/linaleo/start2.htm
Quia – http://www.quia.com/web/
Spanish Language Exercises – http://mid.ursinus.edu/~jarana/Ejercicios/
Traveler – http://www.digitalspace.com/traveler/
Wimba – http://www.wimba.com/

All websites cited in this chapter were verified on 14.09.2002

9

Using internet-based audio-graphic and video conferencing for language teaching and learning

Regine Hampel and Eric Baber

The Open University, U.K. and NetLearn Languages, U.K.

Introduction

In the early stages of CALL, courses were delivered predominantly in a text-based, one-directional mode. With the development of more elaborate web-based technologies so-called "interactive" components were then added. These included activities such as self-marking quizzes and tests along with mechanisms for learners and tutors to interact asynchronously through text – using e-mail, web-based discussion forums and so on. A logical next step, which has been explored and developed for a number of years now, is to use computers and the Internet to support synchronous interaction between learners and teachers. For language teaching, specifically, technologies that allow live voice interaction are of particular interest as they allow the development and practice of speaking and listening skills. This chapter will examine the pedagogies supporting such synchronous computer-mediated communication (CMC), introduce a number of programs that allow such interaction to take place, as well as the tools that these programs offer, and discuss how several institutions have implemented language learning and teaching using synchronous online audio conferencing.

Pedagogical framework

Audio-graphic CMC tools (that is, applications that support tools to allow both audio communication and the manipulation of text and images within the same environment) are relatively new and have not been used (or researched) extensively – apart from a few exceptions (for example, Erben 1999, Hauck & Haezewindt 1999, Hauck et al 1999). However, if we are to provide a valuable experience to both learners and tutors using such tools, it is necessary to take into account the pedagogy appropriate to online language learning and apply it to the practice. There are several existing theories supporting CALL and written forms of CMC (the two terms are used here to point to the distinction often made in the research literature (see, for example, Kern & Warschauer 2000) between CALL as courseware, where the computer is used as a tool and interaction takes place between the learner and the computer, and CMC, which allows the learner to use the computer as a tool for communication with other learners), and these theories can also be applied to synchronous audio-graphic/video conferencing. While theories of second language acquisition (SLA) were appropriate to CALL, a new and wider theoretical framework is required for CMC, a mode of communication which incorporates a sociocultural and multimodal environment. This involves a "move away from the static transmission models for knowledge and skill acquisition that are based on traditional cognitive learning approaches which emphasise learning as an incremental mathematically-facilitated process" (Felix 2002:6).

Second language acquisition theories

Approaches to traditional CALL are typically based on theories of SLA that deal with the language acquisition process by concentrating on aspects like input (Krashen's input hypothesis (see Krashen 1981, 1985), which sees the development of a second language as dependent on the amount of comprehensible input), negotiation of meaning, and comprehensible output (see Swain 1985). As Holliday (1999:186) states:

> *The findings of the research on input and output are applicable to CALL, as computers influence the environment in which learners are exposed to the second language. The design of multimedia language learning products and methods, such as CD-ROMs and computer-mediated communicative language teaching, needs to take input and output factors into account.*

Thus CALL programs should give the students the opportunity – amongst other things – to hear/read comprehensible input, to negotiate meaning, and to produce/write comprehensible output. Similarly, SLA principles can be drawn upon to develop CMC effectively and examine its usefulness, provided that they are opened up to take the social aspects of interaction into account. Such research has been done mainly on written forms of CMC, examining issues such as student participation, turn-taking, interaction (see, for example, Chun 1994, Kern 1995, Kelm 1996, Ortega 1997, Chapelle 2000, Chun & Plass 2000).

Sociocultural theories

The development of CMC (written forms of CMC in the first instance), where groups of learners are able to communicate with each other via their computers, requires the broadening of theories to include the interactive, social aspect of CMC projects, the many-to-many communication scenario.

> *Our desire to understand what is happening when a learner or a group of learners are using a computer has been extended to involve situations where learners collaborate over distance and interact with virtual communities to accomplish creative goals. Research agendas are expanding to include issues*

*of social computing and networked cultures and specific methodologies such
as ethnography and ethnomethodology, designed to further our knowledge in
this area. (Debski and Levy 1999:8)*

Thus, several researchers like Warschauer (1997), Levy (1998) or Debski (1997) have
gone back to Vygotsky's sociocultural theory and applied it to CMC. Vygotsky, who
suggests an "interactionist–dialectical analysis of development" (1978:124) believes
that children learn by being guided through successive 'zones of proximal
development' by interaction both with peers and other people like teachers.

*We propose that an essential feature of learning is that it creates the zone of
proximal development; that is, learning awakens a variety of internal
developmental processes that are able to operate only when the child is
interacting with people in his environment and in cooperation with his peers.
Once these processes are internalized, they become part of the child's
independent developmental achievement. (Vygotsky 1978:90)*

We can also point to situated learning based on Halliday's socially-oriented view of
language and the fact that language has different functions. Language is not just
concerned with construing experience and expressing content – what Halliday calls
the ideational function. It also has an interpersonal function, which is particularly
relevant in the context of oral communication, which means that language consists in
'enacting personal relationships' (Halliday 1993:101), and a textual function, which
refers to the fact that language consists in creating discourse which is situationally
relevant (see Kern & Warschauer 2000). Applying these ideas to second language
learning, Warschauer (1997:487), for example, demands that students be encouraged
'to conduct actively 'meaningful tasks' and solve 'meaningful problems' in an
environment that reflects their own personal interests as well as the multiple purposes
to which their knowledge will be put in the future'.

Sociocultural theories have had a major influence on the development of
constructivism, an approach used by researchers like Rüschoff and Ritter, who
combine theories from cognitive science with a social perspective. Knowledge is seen
as something that must be constructed, not something that can be transferred, and
learning is understood 'as an active, creative, and socially interactive process'
(Rüschoff & Ritter 2001:223). The main focus, therefore, is on the learner, who builds
on already existing knowledge with the help of others. Thus, task-based learning,
using authentic materials in authentic situations, is central to a constructivist learning
approach.

The theories that underpin development and research into the written forms of
CMC, are also helpful for examining the use of audio-graphic and video conferencing
tools. They, too, make many-to-many communication possible, where interaction and
situated learning apply. Thus Vygotsky's and Halliday's theories as well as those
proposed by the constructivists can be usefully employed to inform both the
development of audio-graphic and video conferencing for learning and teaching
purposes and the research into these modes (see also Felix this volume).

Multimodality

The main difference between the new conferencing technologies and more traditional CMC tools is that they no longer limit us to a single (written) mode. Audio adds another layer to CMC, thereby complementing writing by speech. As Halliday stresses, although these two modes represent 'alternative realisations of the meaning potential of language' (Halliday 1986:92), they have different functions and 'impose different grids on experience' (Halliday 1986:93). Writing and speech in CMC can also be supplemented by video and graphics, and it is now possible to offer learners a combination of different modes, which include audio-graphic conferencing, video, text chat, shared writing tools and web access.

Another useful angle on CMC via audio-graphic/video conferencing, therefore, is multimodality, which – according to Kress (2000b:183) – has been developed to try to set 'a new agenda of human semiosis in the domain of communication and representation'. A new approach which necessitates looking at sign systems in order to deal with the demands made by the new electronic technologies (Kress 2000a:158) is called for. Kress & van Leeuwen (2001:20) use 'multimodality' to refer to 'the use of several semiotic modes in the design of a semiotic product or event, together with the particular way in which these modes are combined – they may for instance reinforce each other [...], fulfil complementary roles [...] or be hierarchically ordered'. This mirrors a general development by which writing – which used to be the preferred semiotic mode in Western culture – is becoming a less important mode. This development started with the telephone (which made it possible to use audio communication over a distance), and now we are in a situation where, with the help of computers, we can combine the visual, the verbal and the written, or choose the mode which is most appropriate in a particular situation.

> *Today [...] in the age of digitisation, the different modes have technically become the same at some level of representation, and they can be operated by one multi-skilled person, using one interface, one mode of physical manipulation, so that he or she can ask, at every point: 'Shall I express this with sound or music?', 'Shall I say this visually or verbally?', and so on. Our approach takes its point of departure from this new development, and seeks to provide the element that has so far been missing from the equation: the semiotic rather than the technical element, the question of how this technical possibility can be made to work semiotically, of how we might have, not only a unified and unifying technology, but also a unified and unifying semiotics. (Kress & van Leeuwen 2001:2)*

This is particularly relevant for teaching and learning as a multimodal virtual learning environment is ideally suited to cater for different learning styles. For example, auditory, visual and kinaesthetic types of learners can be offered activities which better match their particular learning styles.

Programs and tools

There are now a number of commercially available programs that allow for synchronous online communication and collaboration. The two perhaps most widely-used products are Microsoft's NetMeeting and CU-See Me by White Pines Software. While NetMeeting was originally developed with business users in mind – allowing colleagues to work on documents and files together at a distance – CU-See Me was initially designed for the educational sector. The products are remarkably similar in that they contain an almost identical set of tools; yet how these tools are employed – for commercial or educational use – depends very much on the needs of the user.

As well as programs developed by corporates for internal training purposes, a number of other programs which are similar to NetMeeting and CU-SeeMe have been produced by educational institutions for use with their staff and students. Examples of these include the software developed by the ReLaTe (Remote Language Teaching) project at the University of Exeter and University College London (UCL), and Lyceum, which was designed by the Open University as a pedagogical tool for teaching and learning. A detailed case study of the ReLaTe project can be found in Buckett & Stringer (2001).

Audio

Both NetMeeting and CU-See Me allow for two-way audio exchanges. This means that if two people log on to the Internet or a local network, start the same program on their computers and establish a connection with each other, they are able to speak to one another. The specification of almost all modern PCs is sufficient to run these applications: a Pentium-class computer with an appropriate soundcard and a microphone and speakers is all that is required. Audio quality is generally very good, i.e. telephone quality or better, even at modem-speeds.

In addition to one-to-one audio, multi-point audio is also possible if an intermediary server is used. In a two-person call the audio travels directly from one computer to the other. In a group-call in which the participants wish to use audio, though, a MCU (Multi-point Conferencing Unit) is necessary. Instead of person A calling person B, all participants in the meeting call the MCU, which receives and redistributes the participants' audio signals. Again, good-quality audio is possible even in a group-call and at modem speeds.

Lyceum works via a system of centralised servers, so that all Lyceum users who are in the same conference are connected to the same voice server and can talk to each other. This voice server forwards all communications to everybody in the room, so that when one person talks, everyone else can hear it. The added advantage of Lyceum is that it allows for 'many-to-many' communication. This means that up to four participants can talk simultaneously, and thus Lyceum imitates a real-life scenario that may be experienced by people communicating face-to-face.

Lyceum also offers the possibility of recording the audio channel so that learners and teachers can review the contents of a tutorial, for example, and reflect on the content in the same way as users of text-based CMC can use logs and conference records for reflection purposes. Tracking is also useful for research purposes. In NetMeeting and CU-See Me the audio conversation cannot be saved without employing additional technology.

Video

Both NetMeeting and CU-See Me also support video in a similar way to audio. Again, with both programs it is possible to have point-to-point video in a two-person call without the need for an intermediary server. For group-video, though, an MCU that supports video is necessary. There is a difference in the way the two programs handle video: with NetMeeting, only one meeting participant can be viewed at a time. Depending on the MCU it may or may not be possible to switch between meeting participants, i.e. being able to view first one member of the meeting, then another. With CU-See Me, on the other hand, it is possible to view all meeting participants at the same time in a collection of smaller windows.

In both cases, the picture quality is largely influenced by the individual user's bandwidth. Video quality at modem speeds tends to be unsatisfactory in that picture/ audio synchronisation cannot be expected and the frame-rate may be quite low leading to a jerky picture. It may also impair the audio quality. The ReLaTe project reports that hardware upgrades were necessary before the video image was deemed to be useful (Matthews 1998:29). At cable or DSL speeds the video picture can be very clear and smooth, though lip synchronisation is still unlikely to be completely satisfactory. In particular when using CU-See Me's multi-video feature the bandwidth use can severely impede overall performance. In the case of both programs, it is possible to turn off individual video channels (in the case of CU-See Me) or to pause/turn off the video-feature completely, thereby allowing the available bandwidth to be used for other tools.

At the moment, video is not being used in online tuition with Lyceum for technical and pedagogical reasons. The technology remains unsatisfactory on most PCs and research has shown that users find the distortion of gestures and expressions very distracting. The technology does not enhance natural group discussion (see, for example, Goodfellow et al 1996) and there is the issue of screen management when dealing with groups of up to 15 participants. While these issues have not been resolved, it is not a priority to further develop the use of video within Lyceum.

Text chat

All three programs include a text-chat feature which allows users to write instant text messages to one another in a communal chat window. These messages can be read by all participants, thereby allowing group communication to take place. At times text chat may be preferable to audio communication. This may be the case when one or more participants have very limited bandwidth, making audio communication difficult or impossible; when wanting to focus on writing or spelling; or when wishing to save a dialogue for later review. One of the constant developmental updates to Lyceum means that it will be possible for text chat to be 'logged' (as is already possible in NetMeeting and CU-See Me). As in the case of recording audio, this offers learners the possibility of reflecting on their performance.

A further function of text chat in NetMeeting and CU-See Me, which users may appreciate, is the ability to send a text message to only one participant. When using group-audio anything that is said by any one member can be heard by all the others;

in some instances this may not be appropriate such as with pair work, teacher correction, and so on. By using the Private text-chat facility a teacher can send a message to a student regarding a mistake they may have made, ask an individual comprehension question of a particular student, and so on. Lyceum does not have such a facility but offers sub-conferences instead; this allows students to work in (small) groups using all the available tools.

Figure 1. Whiteboard in NetMeeting

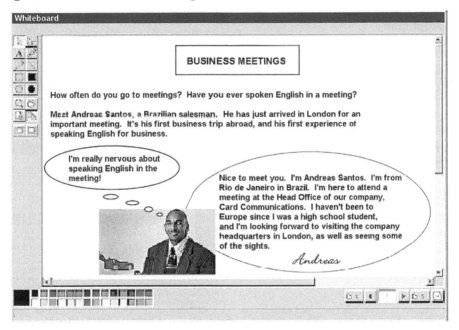

Whiteboard

NetMeeting includes a collaborative whiteboard which allows a teacher to prepare materials for a particular session and then display them on all students' screens at the appropriate time (Figure 1).

The whiteboard can be manipulated by all participants, meaning that once materials have been placed upon the board students can type onto it in order to complete gap-fill sentences, draw lines and other objects onto it in order to perform mix&match activities, and more. The whiteboard can contain text, images, graphs and grids. For educational purposes this is a very useful tool during an online language lesson when used in conjunction with the audio feature.

CU-See Me does not have a whiteboard built in as standard. However, if a computer has both NetMeeting and CU-See Me installed, the NetMeeting whiteboard can be used in a CU-See Me call. One drawback of this is that while there are versions of CU-See Me available for PCs and Apple Macintoshes, NetMeeting is only available for Windows-based machines. While a group meeting is therefore possible in theory between a mix of PC and Apple Macintosh users, only those using PCs and who have additionally installed NetMeeting will be able to use the whiteboard.

In Lyceum, the whiteboard is primarily a graphics tool, where shapes, text, and freehand drawing can be placed on the screen. It is also possible to add images, which can either be grabbed from the screen or loaded from a file.

The whiteboard cannot be used to open Word documents or files prepared or saved in other formats. Materials need to be prepared and saved on the whiteboard itself in order for them to be usable.

Concept map

Lyceum features a collaborative concept map module, which is a tool for writing and for graphically describing concepts and the links between them. Boxes containing text can be added to the display, and arrows can be drawn between them showing how concepts are linked (Figure 2). In language teaching, the concept map can also be used for brainstorming, listing vocabulary or noting down keywords when giving a presentation. Like the whiteboard, the concept map can be seen simultaneously by all participants. However, only one user can key text into a box at one particular time in order to prevent confusion.

Figure 2. Concept map in Lyceum

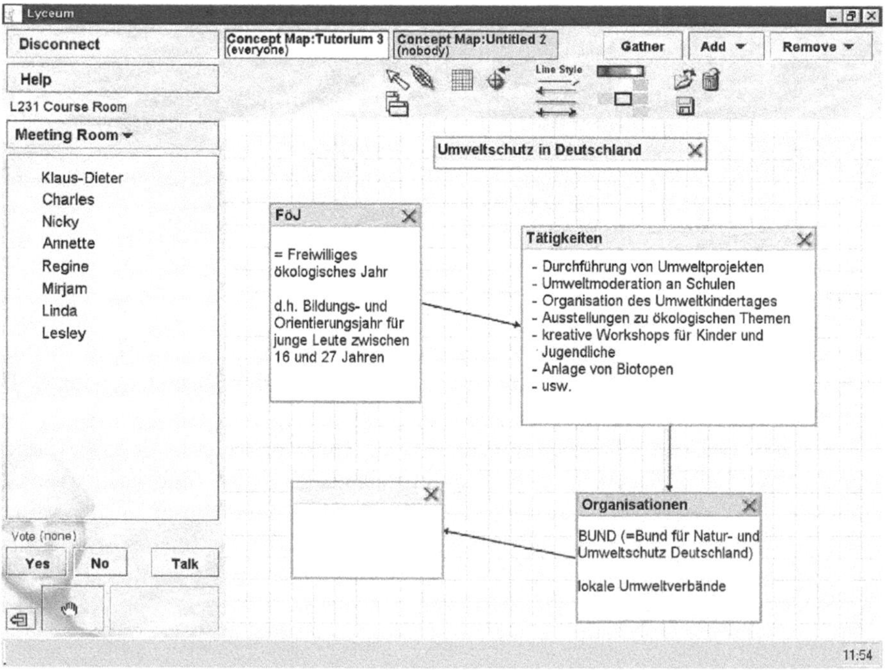

Like the whiteboard the concept map has its own format, and materials saved in this format cannot be opened from within other applications.

Document module

The document module that Lyceum offers is a simple shared word processor. It offers functions like copy and paste and formatting options such as boldface and italics; participants can also choose from different paragraph styles. The module offers added collaborative functions; the 'dispute tags', for example, can be used to mark selected text which needs to be discussed further, and the highlight tool allows the user to highlight specific parts of the text. When dealing with a longer text, it is particularly helpful to be able to use the 'set position' function, which sets the position of the text in each user's shared document window to the same point.

The document module allows multiple users to work on a single document. It can be saved as HTML so that it can be loaded into programs other than Lyceum (for example, Netscape, Internet Explorer, and Microsoft Word).

Sub-grouping facilities

Each conference in Lyceum comprises a number of sub-conferences, or so-called 'rooms', which are accessible at any time during a session (Figure 3).

Figure 3. 'Rooms' in Lyceum (also showing the whiteboard and text chat)

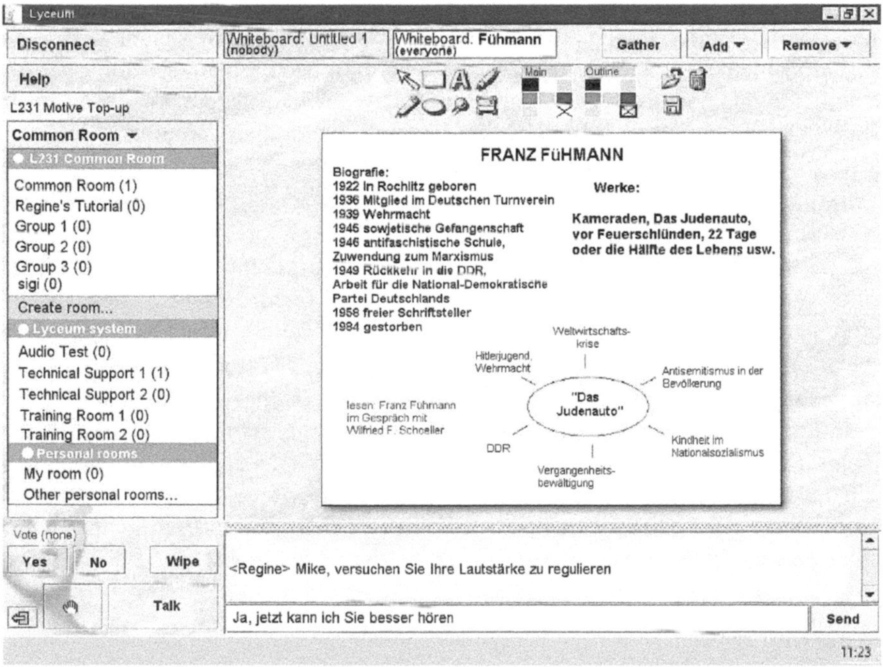

Each person in such a room can talk to as well as hear everybody else in the same room, and students in the room are able to share modules (i.e. whiteboard, concept map, document, and text chat). It is possible to have several modules open simultaneously in the same room, with different participants looking at different modules. Lyceum also has a 'gather' function, which can be used to force all other users in the same room to look at a specific module. It should also be noted that Lyceum follows a democratic model by giving all users the same level of permissions. For the 'gather' function, this means that any user can 'gather' everyone in a room to look at the same module; and any user can alter the contents of a module at will. This gives students the same means of control as the teachers.

While there is a fixed number of existing rooms in each conference, it is nevertheless possible to create as many temporary additional rooms as are needed – for group work during tutorials, for example. Each student also has his or her own personal room, which is only accessible by user name.

File sharing

NetMeeting and CU-See Me both contain a file-sharing feature which allows users to open any document or file on their computer (e.g. Word document, Excel spreadsheet etc.) and share it with other meeting participants. Once shared, the files appear on the other participants' screens. There are some important differences between using the whiteboard and using the file-sharing facility: while the whiteboard can be manipulated by all participants at the same time, the file-sharing functionality enables only one person at a time to have control of the document. File-sharing also throws up a number of security implications in that once a file has been shared with other meeting participants, another participant who has been given control over the document by the document owner could then view, open or delete other files on the owner's computer. The file owner can regain control over their computer at any time during the sharing process but needs to do this manually.

File sharing in these two programs is most useful as a presentation tool, i.e. if one member of the group (the teacher or one of the students) wishes to display a file to all other participants. This can be used for giving mock presentations, displaying the results of a project and so on. For exercises that require two or more people manipulating the material at the same time, the whiteboard tool is preferable to the file-sharing facility.

Lyceum does not offer a file-sharing tool for files which were not created within the program. If participants want to share external information they have first to transfer it into one of the modules. Longer text, for example, can thus be shared with the help of the document module, which approximates to a Word document. For presentation purposes, the whiteboard and concept map can be used as they allow learners and tutors to prepare materials in advance and to display them when needed.

File transfer

NetMeeting and CU-See Me both allow for file-transfers to take place during a call. Transferring a file from one user's computer to another has the same effect as sending it as an e-mail attachment, the main difference being that it takes place within the program. A teacher may choose to transfer a file to the students at the end of the lesson for homework, for instance, or a student may choose to transfer a file s/he has produced as a project to the other students and the teacher in order to discuss it during the rest of the lesson.

Lyceum does not have a direct file transfer tool. It is possible, however, to send a whiteboard, concept map or document as an e-mail attachment.

Supplementary technologies

During or between live online lessons additional technologies may be employed either to enhance the online learning or allow the student to perform language tasks between sessions.

Pre-recorded audio clips

In addition to the live audio offered by conferencing programs, audio can also be used in the form of pre-recorded listening exercises. These can be sent to the student either before or during the lesson via e-mail or using the file transfer facility.

Two popular formats that can be employed are RealAudio and WindowsMedia. Both are widely supported formats, small in size and of very high quality, depending on the connection. The recordings may be made by teachers themselves or by their institution, or may be freely available on the Internet, for example news recordings. In the latter case, students are not sent the actual recordings for copyright reasons; instead, they can be asked to visit the relevant webpage and click on the corresponding link during the lesson. Using news programs also ensures that subject material is topical and up-to-date and thereby of more interest to students. One consideration when using pre-recorded audio clips, however, is the students' computer configurations and hardware set-up. Older computers or soundcards may only be able to play back one audio source at a time; this would mean that the student would not be able to participate in an online call while simultaneously playing back a pre-recorded clip. In this instance, the teacher would be better advised to set the listening task for homework, to be done in between lessons.

Similarly, students may be asked to record themselves using the Sound Recorder which comes built into Windows, and to send these clips to the teacher for evaluation. The teacher can then point out problem areas and make suggestions on how to overcome these. As concentrating on individual needs is not completely satisfactory for students taking part in group lessons, though, many teachers supplement these with occasional one-to-one lessons aimed primarily at pronunciation.

In addition, a number of other tools can be used to supplement synchronous lessons – web-based quizzes and exercises, online tests, and so on.

Implementation of Internet conferencing in language teaching

Different institutions have taken different approaches to the implementation of Internet conferencing in language teaching. Here the following three institutions are considered:

- NetLearn Languages (NLL), an online language school in operation since early 1998, which uses Microsoft NetMeeting for instruction-delivery via the Internet.
- ReLaTe (Remote Language Teaching), a project run jointly by the University of Exeter and UCL, which uses purpose-built software with similar tools to NetMeeting or CU-See Me. Tuition is delivered via the UK-based JANET Mbone Service which utilises multicasting to transmit data and therefore yields faster data-transfer.
- The Open University, which has been using Lyceum, its own Internet-based audio conferencing software, since the beginning of 2002 to deliver online tuition for its distance language courses. Students can also use the program outside scheduled tutorial times in order to meet and practise their communicative skills. Lyceum operates via the Open University's own system of centralised servers.

Student–teacher constellations

There are various scenarios for synchronous online instruction using audio-graphic and video conferencing:
- The teacher is in one location, a group of students at another one. The teacher teaches from his/her computer while the group of students gathers around one computer at a separate location.
- One-to-one – the teacher is in one location while the student is at another one.
- Geographically dispersed groups – the teacher and the students are all in different locations and everyone logs into the conference from their own locations.

The decision as to which mode is most meaningful depends on a number of factors such as financial issues (the lower the student-teacher ratio, for example, the more costly the instruction is likely to be), the type of language learning that is being aimed for, student profile, whether the students are in the same location anyway, and so on.

The first scenario above is most likely to take place when students are in the same location, for example, at a school or university where the language instruction is being offered online because there is no teacher physically present who can teach that particular language. Another situation might be a commercial one, where a number of employees from the same company have similar language needs and levels and can therefore gather in a room set up with the relevant equipment to participate in language learning activities. Although such a scenario was originally envisaged for distance learning and teaching, it is sometimes adopted as a useful teaching mode for participants who occupy the same location. The advantages of such a set-up are that it encourages valuable interaction not only with the teacher via the computer but also among members of the group positioned in front of the computer.

NLL uses the second of the scenarios outlined above and offers predominantly one-to-one lessons aimed at business students, while the Open University follows the model outlined in the third scenario. The Open University's approach was adopted to comply with the growing demands from its students (who all study at a distance from each other and their tutors) for greater flexibility and more practice in speaking and listening skills. After several pilot projects Lyceum was first used as a mainstream tool for a German course at level 2 (post A-level or school leaving exam) and students form tutorials groups which are spread over the UK and continental Western Europe. Online tuition is being progressively introduced for other language courses, including the new beginners' courses. The ReLaTe project also offers lessons to geographically dispersed groups at university level, thereby allowing students to study French, Latin and Portuguese in partial fulfilment of university requirements.

Course design

As in a face-to-face classroom setting, how tuition is structured online depends on the needs of the learner(s). Different deliberations and their outcomes with regards to syllabus design may include the following:

- Is it a one-to-one or a group course? If the course is a sequence of one-to-one lessons, the syllabus is likely to be quite flexible and negotiated with the student or may reflect the outcomes needed for an exam. If it is a group course, the syllabus is likely to be more rigid and pre-defined. ReLaTe courses, for example, have a fixed beginning and end with structure and contents that reflect the need to fulfil university requirements of a language course, while NLL courses are mostly designed on a month-by-month basis depending on the changing needs of the learner.
- Are the online lessons supplemented by any other forms of learning, e.g. classroom-based instruction, self-study work delivered by an online Virtual Learning Environment (VLE) or other form of instruction? At the Open University, for example, online tuition is embedded in a course which also includes course books, audio CDs and videos for self-study.
- Is the course likely to be repeated by another group of students at a future date, or is it a one-time only course? If the former applies, a wider range of learner types may need to be addressed than if the course is designed specifically for a known group of learners.
- Which language skills need addressing, and how are these best approached? While all language skills can be developed and practised in synchronous online tuition, there may be more effective and efficient ways of dealing with certain aspects.

Teaching methods

Although today's online conferencing tools mirror, to a large extent, the tools available in a conventional classroom (whiteboard or audio communication, for example), it is important to consider whether the effect of the virtual environment necessitates that these tools be used differently. Audio communication in an online conference, for example, works differently from a face-to-face session. For sessions with geographically dispersed groups, there are additional issues such as turn-taking or pair and group work, which require special attention in an online setting.

When engaging in group conversation one element that needs to be taken into account is lag, or time delay between when a participant speaks and the other members of the group receive the utterance. Depending on the technology used and the various users' Internet connection speeds, the lag may be between half a second and one second. While this does not rule out group communication, it may take the participants a while to get used to the time lag. It is useful, therefore, to consider mechanisms which ensure the smooth flow of conversation. These include having the next speaker nominated by the current one, introducing a rotation scheme whereby each student in turn gives their opinion, or using some form of visual signalling to indicate that someone would like to speak next. It is possible, for example, to key in a brief message such as 'me' in the text-chat window or to use the visual hand-pointer on the whiteboard in NetMeeting. Lyceum offers the built-in function key of 'raising your hand' to make turn-taking easier.

Learning activities

It is a strength of audio-graphic conferencing that while it generally lends itself to the underlying principles of SLA theories (with the teacher offering a range of stimuli and learning input, and the students being encouraged to produce both written and oral language), in terms of more specific methodology, it supports different approaches. For transmitting factual knowledge, giving presentations or conducting one-to-one tutorials, it is possible to use the transmission approach to language learning. Yet for most other activities a more constructivist approach seems to be more suitable.

Online tuition using audio-graphic/video conferencing lends itself to interactive, collaborative, task-based, student-centred activities based on sociocultural and constructivist theories. This is in line with the gradual shift within CMC that has been identified, moving away

> *from teacher-centred approaches, largely reflected in the explicit teaching of grammar, which exploit the technical potential of the web, to student-centred learning, reflected in meaningful task-based activities, which exploit the new medium's unique potential for authentic learning experiences. (Felix 1999:86)*

In the integrated learning approach of the Open University, the content of the activities used for online tuition relates closely to the content of the course books (which students study at home). The tasks call for a mixture of plenary sessions and group/ pair work during tutorial time and preparation outside scheduled tutorial sessions (with groups meeting in Lyceum and/or communicating via e-mail). The students are encouraged to use authentic web material in order to collate information (based on a vetted selection, but not limited to this – if the students/tutors want to they can search for additional material). In addition, the tasks take into account the multimodal nature of the technology, and encourage students (and teachers) to use different tools – either to suit their particular learning style or to suit the task. Thus one Open University tutor noted:

> *I find different things useful at different times. I personally find the screen-grab very useful as you can import images to back up your work. Concept maps are useful for just writing down points. The text chat becomes handy when the sound system fails.*

Although the activities cover all four language skills, the main focus is on speaking because this is the area in which distance language courses have traditionally provided the least practice.

Activity-types that can be performed with the help of the audio channel in combination with the whiteboard, the concept map or the document module include:

- Writing tasks such as letter-writing, summaries, reports, or articles;
- Ordering activities – a whiteboard may contain the various steps of a production-cycle, but in an incorrect order; students then re-arrange the sentences in the correct order;
- Reading tasks followed by comprehension-check questions;
- Problem-solving activities – students are given different information and – using role-play – have to achieve a goal such as placing an order for a product from a supplier based on price, delivery possibilities etc.;
- Task-based activities, which give the students the opportunity to use authentic material in meaningful tasks, including information-gap activities and role-plays.

Of course, this list is by no means complete. By using their imagination, learners and tutors can participate in a wide variety of learning activities that provide an experience that is different from but no less rich than that of the face-to-face context.

One issue an institution will need to address is the production of materials. In a course with a fixed syllabus as a standard, sequential set of materials can be produced by the institution for use by all its teachers. In a course aimed more specifically at each student's needs, though, this may not be appropriate, and in this case it is likely to be the teacher who prepares each lesson individually. While this will lead to a bank of materials being built up over time, this will probably be more labour-intensive on the part of the teacher than if a standard set of files were produced for shared use.

The role of the teacher

Depending on the student profile and on the institution, the teacher may need to spend time in the first lesson on evaluating the students' needs and negotiating a syllabus accordingly. It is important to start the first lesson with icebreaker and getting-to-know-you activities in order to foster group cohesion. In addition, the teacher may wish to spend some time in the first lesson allowing the students to familiarise themselves with the technology, though this may already have been done by the institution.

As in the face-to-face classroom, the role of the teacher in synchronous online course delivery varies, depending on the pedagogical approach used and the situation s/he finds herself/himself in. Sometimes the online teacher may choose to be somewhat more directive than in the classroom, for example, when waiting for an answer to a question or when dealing with two students speaking simultaneously. In a face-to-face setting the teacher receives and gives visual clues while this is not necessarily the case online. The teacher may also wish to address quiet students more directly; in the classroom there may be visual clues that the student has understood the material and is not actively participating because s/he is naturally shy or quiet. Online, though, there may be different reasons for non-participation – distractions from other sources around the student and lack of comprehension of the subject material are just two examples. By more direct nomination and perhaps more thorough comprehension checking, the teacher is more likely to be able to ascertain the reason for a student's non-participation.

In other settings, however, when following a more constructivist, task-based approach to teaching, for example, the tutor's role is more that of a facilitator, a co-worker or a participant in the student's learning (see, for example, Fox 1998). When evaluating a project using online audio conferencing, Hauck & Haezewindt (1999:50, see also Hauck et al 1999) found that 'the tutor turned gradually into a manager of learning resources and an organizer of learning events'. Similarly, one of the tutors involved in Lyceum use reported that the students had more control over the learning situation and that her approach had generally been 'more hands-off'.

Opportunities and pitfalls of online learning

One recurring problem in online teaching is that of retention rate. The retention rate for online, self-study courses is not high, often as a result of external pressures, lack of motivation, lack of interaction with peers or the teacher, or other reasons. For synchronous online courses taught by NetLearn Languages meanwhile, the attrition rate is substantially lower: figures appear to be around 10% as opposed to 40–50% in asynchronous courses. There seem to be several reasons for higher completion rates in these synchronous courses:

- Because lessons are scheduled to take place at a certain time, students plan it into their day and then take part in the lessons. A student taking part in an asynchronous course may attempt to study in a less structured manner and may find it more difficult to be self-motivated enough to set aside the required study-time.
- Frequent, or regular, direct contact with the teacher and other students appears to motivate students. Because they are actively using the language with other students – some of whom may be in different parts of the world – they are more likely to appreciate the progress they are making and use the new language they have learned.
- In a group-scenario students may be able to contact each other outside scheduled sessions, either synchronously or asynchronously. If the students have never physically met there are likely to be natural information gaps about their lifestyles, jobs, countries and so on, making it fascinating for students to exchange real information.
- The online environment as a whole appears to be highly motivational to students who in a classroom situation are likely to be shy or quiet. Especially when no webcameras are being used students are likely to actively participate in online sessions. The fear of making a mistake and looking foolish in front of a group of peers seems to be drastically reduced or removed completely; except in situations where a group of students are taking part in a lesson from a shared computer, students are highly unlikely to ever meet each other face-to-face and barriers are therefore substantially lower.

At the Open University, the situation is slightly different. Open University language courses are not offered solely online and participation in tutorials is not a prerequisite to the study of a language at a distance. Yet it is as important to offer first-rate tuition and when online tuition was first used in a level 2 German course, the effects of its introduction were carefully researched. Thus a range of benefits and challenges were identified.

At the beginning of the course, many students complained about technical difficulties, like fluctuating audio levels and poor audio quality, or being disconnected in the middle of a tutorial. Yet it was also found that at the same time students seem quite reluctant to seek help when problems occur. Because of the lack of body language, awkward silences occur and sometimes online tutorials seem less spontaneous. It may also make it more difficult for shy people to participate. Research about the effect of written forms of CMC on the participation of shy students suggests that they contribute more frequently than in the face-to-face classroom (see Chun 1994, Kern 1995, Warschauer 1996), but it is unclear as yet whether this research is transferable to the effect of audio(graphic) conferencing on learner contribution. Several students complained about the complexity not of the system but of having to do too many things at once (using the mouse, speaking, typing on the keyboard), and some of the tools were seen as laborious and time-consuming (especially for those students whose typing skills are not very good). As one tutor noted, the technical challenge adds to the difficulties encountered in speaking a foreign language. Some students and tutors noted the need for more practice with Lyceum (which of course they could do something about themselves but presumably this is a time issue). A general point was the fact that students found it difficult to find the time to prepare for the sessions and that tutors then had to adapt the activities. Another complaint was that not all students participated on all occasions (as tutorials are not compulsory) and therefore the groups were sometimes too small to do the collaborative tasks justice.

Despite the problems (some of which were resolved by further development of the system, especially with respect to audio quality), both students and tutors identified a number of benefits of Lyceum. Here they pointed to authentic communication with other learners over a distance and the opportunity to practise and improve the target language. Several mentioned the fact that Lyceum is very useful for working in small groups. Some liked the idea of sharing texts. One student summarised the experience as follows:

> *I think it [Lyceum] is brilliant and it helps to overcome a major shortcoming of the Open University language courses, namely the opportunity to speak the language. The fact that we can see texts, send texts to each other is [also] fantastic.*

Other reported advantages include an increase in computer literacy and the fact that students become more familiar with the Internet and are able to search for information for assignments more easily. Tutors commented on the usefulness of online tuition for oral interaction and work in small groups. Some were impressed by the way students used the different tools, e.g. to make presentations using several of them, and observed that users found Lyceum exciting and stimulating. As there is no need to travel, Lyceum enables fairly regular attendance at tutorials, and of course students can meet independently at any time they wish. The tutors also appreciated the fact that the students exerted more control over their learning.

The experience at the Open University with synchronous audio tuition has thus shown that using new technology poses a range of challenges. To a certain extent this is due to difficulties with the technology, difficulties which still persist and which, in the main, are related to ISPs and bandwidth. At the same time, students have to get used to the new medium, the tools and the virtual environment, that is, a learning situation which is different from that of a face-to-face classroom. Once they feel comfortable with this situation, however, participants in online tuition appreciate the opportunity not only to practice their speaking and listening skills but also to communicate at a distance using authentic material in a multimodal environment. This coincides with the positive experience that learners at NetLearn Languages have reported.

Conclusion

Synchronous online teaching using audio conferencing (possibly complemented by video) in some ways mirrors face-to-face classroom-based instruction. Many of the underlying methodological considerations are the same when designing a synchronous online course as when designing a face-to-face one, and many teaching methods, which have proven to be useful in a traditional classroom, can be transferred to the virtual environment. Yet there are also major differences. Audio-graphic and video conferences feature tools which are not available in a traditional classroom (web access, pasting and sharing web images, text chat, or tools with which a group can write a text together). The online teacher will thus need to be comfortable designing electronic materials for use on a whiteboard, a concept map or a document module. This does not, necessarily, require any knowledge of HTML or website design (any teacher familiar with a word-processing program is likely to be able to produce materials to be used in a synchronous online lesson), but it does require an understanding of basic accessibility and usability issues as they relate to electronic learning materials.

The different challenges posed by teaching in a virtual environment using audio-graphic/video conferencing also have to be taken into account. We must not forget the fact that the technology is not just a neutral delivery tool but has an impact on teaching and learning. As Levy (2000: 190; see also Warschauer 1998: 84–85) states:

> For the CALL researcher, technology always makes a difference; the technology is never transparent or inconsequential. [...] At the very least, in CALL research, the particular technology in use is taken to affect the language produced, the learning and teaching strategies, the learner attitudes and the learning process. Beyond that, technology may affect the ultimate goals of learners, the nature of the learning environment, teacher education and what it means to be competent in a language.

The challenge, therefore, is to make sure that the effect of the technology is a positive one. The aim is to improve students' language skills while giving them a positive experience in the virtual environment. While the technology currently in use is still being developed and has not yet been perfected, the increase in spread of broadband

Internet connectivity around the world is doing much to make synchronous online instruction a viable alternative to face-to-face, classroom-based instruction. It is important to remember, however, that in and of itself technology is unlikely to make for good teaching. If technology is used productively and imaginatively by an experienced teacher as a resource, it can, however, greatly enhance the teaching and learning process. It is vital therefore that we think about the pedagogies used in online teaching and learning to ensure that they match both student and institutional needs.

References

Buckett, J. & Stringer, G. (2001). A case study in videoconferencing language teaching. In A. Chambers & G. Davies (Eds.), *ICT and Language Learning: A European Perspective,* 167-173. Lisse: Swets & Zeitlinger.

Chapelle, C. A. (2000). Is network-based learning CALL? In M. Warschauer & R. Kern (Eds.), *Network-based Language Teaching: Concepts and Practice*, 204–228. Cambridge: Cambridge University Press.

Chun, D. M. (1994). Using computer networking to facilitate the acquisition of interactive competence. *System*, 22 (1), 17–31.

Chun, D. M. & Plass, J. L. (2000). Networked multimedia environments for second language acquisition. In M. Warschauer & R. Kern (Eds.), *Network-based Language Teaching: Concepts and Practice,* 151–170. Cambridge: Cambridge University Press.

Debski, R. (1997). Support of creativity and collaboration in the language classroom: A new role for technology. In R. Debski, J. Gassin & M. Smith (Eds.), *Language Learning through Social Computing,* 39–65. Melbourne: University of Melbourne.

Debski, R. & Levy, M. (1999). Introduction. In: R. Debski & M. Levy (Eds.), *WORLDCALL: Global Perspectives on Computer-Assisted Language Learning,* 8–10. Lisse: Swets & Zeitlinger.

Erben, T. (1999). Constructing Learning in a Virtual Immersion Bath: LOTE Teacher Education through Audiographics. In R. Debski and M. Levy (Eds.), *WORLDCALL: Global Perspectives on Computer-Assisted Language Learning,* 229–248. Swets & Zeitlinger, Lisse.

Felix, U. (1999). Web-Based Language Learning: A Window to the Authentic World. In R. Debski & M. Levy (Eds.), *WORLDCALL: Global Perspectives on Computer-Assisted Language Learning,* 85–98. Lisse: Swets & Zeitlinger.

Felix, U. (2002). The Web as vehicle for constructivist approaches in language teaching. *ReCALL*, 14 (1), 2–15.

Fox, M. (1998). Breaking down the distance barriers: perceptions and practice in technology-mediated distance language acquisition. *ReCALL*, 10 (1), 59–67.

Goodfellow, R., Jefferys, I., Miles, T. & Shirra, T. (1996). Face-to-face language learning at a distance? A study of a videoconference try-out. *ReCALL*, 8 (2), 5–16.

Halliday, M.A.K. (1986). *Spoken and Written Language*. Victoria: Deakin University Press.

Halliday, M.A.K. (1993). Towards a Language-Based Theory of Learning. *Linguistics and Education*, 5 (2), 93-116.

Hauck, M. & Haezewindt, B. (1999). Adding a new perspective to distance (language) learning and teaching – the tutor's perspective. *ReCALL*, 11 (2), 46–54.

Hauck, M., Hewer, S. & Shield, L. (1999). Online Media for Language Learning. *CILT CALL Report*. UK: CILT.

Holliday, L. (1999). Theory and Research: Input, Interaction, and CALL. In J. Egbert & E. Hanson-Smith (Eds.), *CALL Environments: Research, Practice, and Critical Issues,*181–188. Alexandria: TESOL.

Kelm, O. R. (1996). The Application of Computer Networking in Foreign Language Education: Focusing on Principles of Second Language Acquisition. In M. Warschauer (Ed.), *Telecollaboration in Foreign Language Learning: Proceedings of the Hawaii Symposium,* 19–28. Honolulu: University of Hawaii.

Kern, R. (1995). Restructuring Classroom Interaction with Networked Computers: Effects on Quantity and Characteristics of Language Production. *The Modern Language Journal*, 79 (4), 457–476.

Kern, R. & Warschauer, M. (2000). Introduction: Theory and practice of network-based language teaching. In M. Warschauer & R. Kern (Eds.), *Network-based Language Teaching: Concepts and Practice,* 1–19. Cambridge: Cambridge University Press.

Krashen, S. (1981). *Second Language Acquisition and Second Language Learning.* Oxford: Pergamon.

Krashen, S. (1985). *The Input Hypothesis: Issues and Implications.* London: Longman.

Kress, G. (2000a). Design and Transformation: New theories of meaning. In B. Cope & M. Kalantzis (Eds.), *Multiliteracies: Literacy learning and the design of social futures,* 153–161. London: Routledge.

Kress, G. (2000b). Multimodality. In B. Cope & M. Kalantzis (Eds.), *Multiliteracies: Literacy learning and the design of social futures,* 182–202. London: Routledge.

Kress, G. & van Leeuwen, T. (2001). *Multimodal Discourse: The modes and Media of contemporary communication.* London: Arnold.

Levy, M. (1998). Two conceptions of learning and their implication for CALL at the tertiary level. *ReCALL*, 10 (1), 86–94.

Levy, M. (2000). Scope, goals and methods in CALL research: questions of coherence and autonomy. *ReCALL*, 12 (2), 170–195.

Matthews, E. (1998). Language learning using multimedia conferencing: the ReLaTe project. *ReCALL*, 10 (2), 25–32.

Ortega, L. (1997). Processes and outcomes in networked classroom interaction: Defining the research agenda for L2 computer-assisted classroom discussion. *Language Learning & Technology*, 1 (1), 82–93.

Rüschoff, B. & Ritter, M. (2001). Technology-Enhanced Language Learning: Construction of Knowledge and Template-Based Learning in the Foreign Language Classroom. *Computer Assisted Language Learning (Special issue: Theory & Practice: The German Perspective)*,14 (3–4), 219–232.

Swain, M. (1985). Communicative competence: some roles of comprehensible input and comprehensible output in its development. In S. Gass and C. Madden (Eds.), *Input in Second Language Acquisition,* 235–253. Rowley/Mass.: Newbury House.

Vygotsky, L. S. (1978). *Mind in Society: The Development of Higher Psychological Processes.* Cambridge, MA: Harvard University Press.

Warschauer, M. (1996). Comparing face-to-face and electronic discussion in the second language classroom. *CALICO Journal*, 13 (2–3), 7–26.

Warschauer, M. (1997). Computer-Mediated Collaborative Learning: Theory and Practice. *The Modern Language Journal*, 81 (4), 470–481.

Warschauer, M. (1998). Online Learning in Sociocultural Contexts. *Anthropology & Education Quarterly*, 29 (1), 68–88.

Websites

CU-See Me – http://www.cuseeme.com
Lyceum – http://www.open.ac.uk
NetLearn Languages – http://www.netlearnlanguages.com
NetMeeting – http://www.microsoft.com/windows/netmeeting
RealAudio – http://www.real.com
ReLaTe – http://www.ex.ac.uk/pallas/relate/
WindowsMedia – http://www.microsoft.com/windows/windowsmedia

All websites in this chapter were verified on 18.09.2002

10

Perspectives on offline and online training initiatives

Graham Davies, Thames Valley University, U.K.

Introduction

Training is an area that is commonly neglected both in the business community and in the education sector. The training budget is often the first to be cut in times of financial crisis, and there is a tendency for administrators to regard training as a one-off rather than an ongoing process – which is a general cause of the ineffectiveness of training but particularly so in the case of training in the implementation of new technologies. One only has to look back at the failure of earlier technologies to make a significant impact on teaching and learning to see what can go wrong. It was not the technologies that were at fault; it was the lack of appropriate training for teachers in making the best use of them (Davies 1997:28).

Failure to provide adequate training in the use of new technologies leaves the door wide open for sceptics and cynics to mount their attacks. Oppenheimer (1997) is a typical example, arguing that, because earlier technologies failed to make an impact on education, history is likely to repeat itself with the advent of computer technology. He also argues that providing training is one of the more expensive long-term responsibilities – along with upgrading software and network maintenance – and one that schools can ill afford. This is rather like saying one should not buy a car because it is costly to learn how to drive it, to fill it up with petrol at regular intervals, and to pay a mechanic to maintain it. There is, of course, a fundamental difference between learning how to drive a car and learning how to use a computer. Automobile technology is relatively stable, and once one has learned how to drive there are few new things that one has to learn as the years pass by. Computer technology changes daily, however, and one often feels caught up in a process of 'dynamic obsolescence'. (Davies 1997:27)

In the world of education there is considerable pressure on teachers to keep up with the latest technology, and those who do not have state-of-the-art hardware are often made to feel inferior, but throwing new hardware at a problem is not the solution:

Decision making on policies and programs to promote ICT use often relies too much on absolute numbers rather than qualitative aspects of connectivity. There is a tendency to believe that more is better – more Internet users, more computers, more computer labs. However, a focus on extending ICT coverage

without complementary training or content can dilute users' experience with
ICTs, leaving users with poor quality access or turning them off from the
technology completely. (Kirkman et al 2002:23–24)

The phrase that stands out here is *complementary training or content*. The provision
of *complementary training* has to go hand in hand with the provision of equipment and
access, and the *content* has to be relevant to the intended users.

There are two key issues that must be taken into consideration when determining
the relevance of the content of an ICT training course for language teachers:
- The content of the course must reflect the state of development of the
  technology to which the trainees have access or to which they are likely to
  have access in the near future.
- The content of the course has to be subject-specific.

Considering the first of these two issues, Kirkman et al (2002) report on the
compilation of the Networked Readiness Index (NRI), in which 75 countries
representing 80% of the world's population are ranked in terms of their potential to
exploit ICT – not specifically in education but in general terms. The NRI ranks the 75
countries according to their capacity to take advantage of ICT networks, bearing in
mind key enabling factors as well as technological factors: for example the business
and economic environment, social policy, and the educational system. Higher ranked
countries have more highly developed ICT networks and greater potential to exploit
the capacity of those networks.

Interestingly, and perhaps not surprisingly, there is a strong correlation between
the NRI and the countries of origin of visitors to the *ICT4LT Website* (Davies 2002:8–
9). The latter is an online training resource for language teachers, more fully discussed
later. Fifteen out of the top twenty countries in the NRI also appear in the top 20
countries of origin of visitors to the *ICT4LT Website* – statistics based on data
collected up to April 2002. Table 1 shows the respective rankings. Iceland, Norway,
Singapore, Taiwan and Korea do not appear in the first 20 of the ICT4LT column,
although each of these countries, with the exception of Iceland, has logged a
significant number of visits to the *ICT4LT Website* on occasions. There is also a strong
correlation between countries at the bottom of both lists – not shown in the table here.
China, the whole of the African continent (with the exception of South Africa), and
the whole of the Indian subcontinent appear close to the bottom of both lists.

Table 1. NRI and countries of origin of ICT4LT Website visitors

Ranking	NRI	ICT4LT Website visitors
1	USA	UK
2	Iceland	Italy
3	Finland	Finland
4	Sweden	Spain
5	Norway	Belgium
6	Netherlands	USA
7	Denmark	Denmark

Ranking	NRI	ICT4LT Website visitors
8	Singapore	France
9	Austria	Sweden
10	UK	Canada
11	New Zealand	Netherlands
12	Canada	Germany
13	Hong Kong	Australia
14	Australia	New Zealand
15	Taiwan	Ireland
16	Switzerland	Switzerland
17	Germany	Japan
18	Belgium	Austria
19	Ireland	Hong Kong
20	Korea	Poland

The above table indicates that, in the short term, only a relatively small group of rich and liberal countries concentrated in North America, Europe, the Antipodes and the Far East (excluding mainland China) are likely to benefit from online delivery of learning or training materials. The World Wide Web is not yet accessible worldwide. As a matter of high priority, efforts therefore have to be made to reach out to those parts of the world that are currently underserved in the area of information and communications technology. This is the main aim of *WorldCALL*, an international consortium of professional associations devoted to promoting language learning and teaching with the aid of new technologies.

An example of an outreaching initiative that has enjoyed considerable success is the European Commission's TEMPUS/PHARE/TACIS program, which was set up in 1990 in order to help Central and Eastern European countries rebuild their higher education infrastructures in the post-Communist era. One of the strands of the program provided funding for setting up centres in selected educational institutions for delivering courses in ICT for language teachers, for example the EECALL Centre at *Dániel Berzsenyi College*, Hungary. Two categories of trainees were targeted at the EECALL Centre: (i) students undergoing initial teacher training, (ii) teachers already in service. Most of the in-service trainees who participated in the EECALL Centre courses in the early 1990s did not have access to computers that were capable of running the latest CALL software, so the courses focused initially on using generic applications, authoring templates and concordancers. Multimedia CD-ROMs were demonstrated, but accessing the Internet was out of the question as the local telecommunications networks were unreliable. Hungary was quick to update its antiquated networks, however, and by the mid-1990s the Internet featured prominently in training courses. The provision of discrete outside help, a commitment on the part of the national government to renovate its infrastructures, and the delivery of training at a pace reflecting the state of local technological development proved to be a formula for success. In 1996 Dániel Berzsenyi College hosted the *EUROCALL* annual conference, and Hungary is now ranked No. 30 in the NRI.

Considering the second of the two issues identified above, namely the provision of subject-specific content in an ICT course for language teachers, an introductory course might consist of four discrete topics:

- Using generic software in language learning and teaching
- Exploiting the World Wide Web
- CALL pedagogy and methodology
- CALL authoring tools

One topic per day could be covered in a face-to-face introductory course, followed up by further study of online materials. Trainees could also communicate with one another and with tutors via a bulletin board or a discussion list. The following four sections of this chapter look at each of these topics in more detail. In this discussion we draw extensively on our recent experience in setting up and managing the ICT4LT project and participating in the TALLENT (Teaching and Learning Languages Enhanced by New Technologies) project, both of which have resulted in the production of a large body of online and offline ICT training materials for language teachers. In addition, we draw on feedback from trainers and trainees that has emerged from two UK training initiatives: the NOF (New Opportunities Fund), a general ICT training initiative for teachers, and CILT-NOF, an ICT training course for language teachers that comprises a substantial online element. All four initiatives are described in detail in the concluding sections of the chapter.

Using generic software in language learning and teaching

As a first step, it is essential to familiarise language teachers with generic software. Such software is likely to have been supplied along with the hardware to which they have access, and it is a prerequisite for subsequent CALL-specific training as many of the basic operations teachers learn will be transferable. Generic software includes:

- the Windows operating system and accessories
- a word processor, e.g. *Word*
- presentation software, e.g. *PowerPoint*
- a spreadsheet, e.g. *Excel*

A problem that is frequently encountered by trainers is the varying levels of knowledge of generic software that language teachers come equipped with when they join a training course in ICT. They often have an insufficient knowledge of Windows, including basic operations such as copying and pasting between two open applications. Their knowledge of word processors is often patchy, which is one of the biggest obstacles to efficient progress. For this reason trainees should be invited to assess their own strengths and weaknesses in the use of generic software – and this could be extended to cover CALL-specific software too. Table 2 shows an extract from a 'can do' document located at the *ICT4LT Website*. Each trainee is asked to indicate his/her level of knowledge of each application as (1) Basic, (2) Intermediate or (3) Advanced, and then to check the 'can do' list. This enables the trainer to identify gaps in the trainees' knowledge and also helps the trainees assess the development of their skills and understanding as they progress through the course.

Table 2. Extract from a 'can do' list: word processor

I would describe my ability to use a word processor as (1-3)

Now indicate what you can do: tick for "yes", leave blank for "no". I can:	
Start a word processor	
Exit a word processor	
Open a new document	
Type at a reasonable speed	
Set paper size (e.g. A4) and margins	
Set paper orientation to portrait or landscape	
Insert page numbers into a document	
Insert headers and footers into a document	
Save a document that I have typed	
Print a document	
Open a document that I have previously saved	
Amend/add to a document that I have previously saved	
Save a document that I have amended or added to	
List continues…	

The 'can do' lists may also include 'I understand' statements, for example:

Table 3. Extract from an 'I understand' list: web browser

Essential things that I understand

I understand that a new window sometimes opens when I link from one website to another.	
I understand how frame-based websites work.	
I understand that 'cookies' need to be activated while I am browsing certain websites.	

Training in the use of generic packages is often delivered as part of a general ICT course that trainees follow prior to participating in a CALL-specific course. Having followed such a course, trainees may be able to carry out basic operations but they may be unaware of the operations that language teachers consider important, for example typing accented and umlauted characters and enhancing a *Word* document with a picture or a sound file in order to create an electronic worksheet for their students. It can be argued that CALL specialists rather than general ICT specialists are better equipped to deliver courses in the use of generic packages as they are more likely to understand the needs of language teachers and engage the trainees in contextualised tasks. For example, a German language teacher can be taught the drag-and-drop operation in a word processor by being asked to rearrange the words in the following sentence, beginning with 'jeden Tag':

Mein Vater fährt jeden Tag mit dem Auto in die Stadt.

This kind of operation is equally useful for learners, who can be introduced to the rules of word order in a German sentence at the same time as they improve their word processing skills.

A word-processor can be used to create many different types of CALL activities, for example:

- 'word snake' activities
- sorting and categorising activities
- drag-and-drop and gap-filling activities
- vocabulary building activities, using the *Word* thesaurus

A word processor and a spreadsheet can also be used as an aid to marking students' work and recording their grades:

- Students' work submitted in *Word* can be marked using the *Track Changes* and *Reviewing* options.
- *Excel* can be used to set up electronic mark books.

In recent months there has been an increasing demand for training in the use of *PowerPoint*. Large numbers of schools in the UK are now installing interactive electronic whiteboards in their classrooms for whole-class teaching, and *PowerPoint* is one of the tools that works well in conjunction with this new technology. Electronic whiteboards add a new dimension to whole-class teaching, but a considerable investment of time in training teachers to make the best use of them is essential. Face-to-face training is more effective, but videocassettes of teachers using electronic whiteboards for teaching foreign languages are available, and Module 1.3 at the *ICT4LT Website* contains tips on how to use them and instructions on enhancing *PowerPoint* presentations with pictures and sound. There are also examples of *PowerPoint* presentations that can be downloaded, as well as links to sites containing further examples and ideas. It is likely that this will be a growth area in the future.

Exploiting the World Wide Web

Once they have received basic training in using the generic software packages described above, language teachers should be ready to make use of a browser – which can also be considered a generic software package these days – and explore language resources on the web. The 'can do' document referred to in the previous section also includes 'can do' lists for a browser and for email software, as well as a number of 'I understand' statements concerning plug-ins, copyright, and the importance of keeping anti-virus software up to date and using a firewall while connected to the Internet.

Workshops on exploiting the web typically begin with an atmosphere of excitement as the participants skip from site to site all around the world, followed by the gradual realisation that the wealth of materials available means that careful planning ahead and setting of structured meaningful tasks for students is the key to success. Trainees learn that unstructured browsing can be entertaining – and often informative – but it does not necessarily lead to successful language learning:

> *It is therefore important that teachers learn how to find relevant materials on the web and, above all, how to integrate them into their teaching. (Vogel 2001)*

Most trainees quickly get used to the basic operations offered by a browser, and there are a number of websites offering high-quality tutorial materials that enable them to make further progress on their own, for example:

- The *Web Literacy* tutorial materials, written by Bernard Moro and located at the website of the Council of Europe's European Centre for Modern Languages.
- The *Web Skills for Language Learners* tutorial materials, written by Charlie Mansfield & Tony McNeil and located at the website of the WELL Project.

Training language teachers how to locate relevant materials is an important aspect of web training. Training in the use of search engines is essential. Categorised lists of useful websites can also be consulted, thereby saving a good deal of searching and browsing time. Felix (2001:187ff.) categorises and annotates around 600 websites, and there are numerous website links in the Resources Centre at the *ICT4LT Website*.

Once they have located appropriate materials on the web, teachers need to be trained in how to exploit them – which is an aspect of CALL methodology and considered in the following section. Some websites are designed for autonomous language learners, for example *BBC Languages*, and others contain collections of ready-made materials for classroom use, for example *Bonjour*. There are also numerous sites containing authentic materials that lend themselves to exploitation in the languages classroom. Windeatt et al (2000) have compiled a substantial collection of classroom activities for learners of English as a Foreign Language that exploit authentic materials on the web. While the activities are aimed primarily at students, they can also be used very effectively in training teachers. The activities all require the use of a browser, often in combination with a word processor and other software tools. Gitsaki & Taylor (1999) have compiled a similar collection of colourfully illustrated activities for EFL learners. Most of the activities could be adapted for languages other than English.

Learning how to evaluate a website is a key aspect of teacher training. It is also important that teachers realise the importance of respecting copyright when downloading or copying materials from another website. The following basic advice should be given:

- Downloaded materials should never be given or lent to colleagues in other institutions.
- Downloaded materials should never be reproduced in printed publications or at other websites without the express permission of the copyright holder.
- It cannot be assumed that because material is publicly available on the web teachers can do whatever they like with it.
- There is no truth in the myth that copyright can be disregarded if the materials in question are for educational use.

Finally, it is important that language teachers are made aware of some of the problems associated with using the web. The web is a key source of viruses, for example, and trainees can often be observed stumbling accidentally across websites containing offensive materials that they would not wish their students – particularly younger students – to be exposed to. An awareness of web security is therefore an essential part of a training course.

CALL pedagogy and methodology

The *ICT4LT Website* contains a module on CALL methodology which has consistently been the least visited module since it first appeared. It may be that site visitors are put off by the abstract-sounding title of the module, even though it contains sound practical advice on structuring the learning environment and providing and delivering language learning resources. Littlemore (2002:4–5) reports that the theoretical aspects of a TALLENT course (referred to in detail later in this chapter) did not prove popular with trainees. Successful implementation of CALL is more likely to follow if pedagogical and methodological issues are given due consideration, so why is there this apparent lack of interest? Apart from the negative reactions that many language teachers have towards anything that sounds abstract and theoretical, there is also a possibility that the attractions of the technology itself cause pedagogical and methodological issues to be pushed into the background.

Training teachers to spot examples of technology driving – or restricting – the pedagogy is crucial. This is particularly important with regard to multimedia CD-ROMs and web-based language learning materials. Far too many multimedia CD-ROMs are little more than impressive presentations, with the interaction being limited to a handful of multiple-choice and drag-and-drop exercises. Web-based language-learning materials often suffer from the same shortcomings. It appears that the lessons that were learned about input analysis, feedback and branching way back in the 1980s have been forgotten. As Levy (1997:ix) points out, the concepts and principles that informed designers of earlier CALL software 'do not necessarily become obsolete when the computer that is used to run them is retired'. Many valuable lessons can be learned from the past (Davies 1997).

Pedagogy and methodology need not be dealt with in an abstract way. For example, trainees can be presented with a range of different types of CALL software and asked to evaluate them. This is difficult to do in a distance-learning situation as it implies that the trainees have access to a software library. There are, however, an increasing number of websites offering CALL activities that previously were only available on CD-ROM or DVD-ROM. Evaluation criteria include:

- functionality – ease of navigation, types of interaction, flexibility
- media content – appropriateness of mix of text, images, sound and video
- quality of linguistic and cultural content
- relevance to different programs of study
- expected learning outcomes

Alternatively, trainees might be invited to bear in mind the following questions when examining a software package or website:

- Is the level of language that the program/website offers clearly indicated?
- Is the user interface easy to understand? For example, are there ambiguous icons that cause confusion?
- Is it easy to navigate through the program/website? Is it clear which point the learner has reached?
- What kind of feedback is the learner offered if he/she gets something wrong? Is the feedback intrinsic or extrinsic?
- If the learner gets something right without understanding why, can he/she seek an explanation?
- Can the learner seek help, e.g. on grammar, vocabulary, pronunciation, cultural content?
- Does the program/website branch to remedial routines?
- Can the learner easily quit something that is beyond his/her ability?
- Does the learner have to mentally process the language that he/she sees and hears? Or does the program/website offer a range of point-and-click activities that can be worked through with the minimum of understanding?
- If the program/website includes pictures, are they (a) relevant, (b) an aid to understanding?
- If the program/website includes sound recordings, are they of an adequate standard? Are they (a) relevant, (b) an aid to understanding? Is there a good mix of male and female voices and regional variations?
- Can the learner record his/her own voice? Can the learner hear the playback clearly?
- Does the program/website make use of Automatic Speech Recognition (ASR)? Is it effective?
- If the program/website includes video sequences, are they of an adequate standard? Are they (a) relevant, (b) an aid to understanding?
- Does the program/website include scoring? Does the scoring system make sense? Does it encourage the learner?

Warschauer's three phases of CALL illustrating the development of different ways in which the computer has been used in language learning and teaching, may also be useful as a starting point for training teachers to understand pedagogical and methodological issues. The three phases that Warschauer (1996) distinguishes can be summarised as:

- Behaviouristic: In this phase the computer plays the role of tutor, serving mainly as a vehicle for delivering instructional materials to the learner.
- Communicative: In this phase the computer is used for skill practice, but in a non-drill format and with a greater degree of student choice, control and interaction. This phase also includes (i) using the computer to stimulate discussion, writing or critical thinking (e.g. using programs such as *Sim City*), and (ii) using the computer as a tool – examples include word processors, spelling and grammar checkers, and concordancers.
- Integrative: This phase is marked by the introduction of two important innovations: (i) multimedia, (ii) the Internet. The main advantage of multimedia packages is that they enable reading, writing, speaking and

listening to be combined in a single activity, with the learner exercising a high degree of control over the path that he/she follows through the learning materials. The Internet has numerous advantages, building on multimedia technology and in addition enabling both asynchronous and synchronous communication between learners and teachers. A range of new tasks is made possible, e.g. web searches, web concordancing, and collaborative writing.

Both multimedia and online materials have improved in recent years, embodying a constructivist rather than behaviouristic approach to CALL, but as Felix (2001:191) points out:

While the web is providing an increasingly rich shared resource to CALL practitioners, the often alluded to radical rethinking of the teaching approach still has a long way to go. (Felix 2001:191)

Many software developers continue to offer no more than drill-and-practice materials in a new guise rather than seeking new approaches to language learning and teaching that are opened up through new technologies. Levy (1997:118ff.) analyses the results of a comprehensive CALL Survey that he carried out among authors of CALL materials in order to determine what kinds of conceptual frameworks lay behind their work. The CALL Survey was concluded in early 1991 – before multimedia CALL was widely available, and before the advent of the web. Although there was strong support among Levy's respondents for the communicative approach to language teaching and task-based learning, a substantial number also favoured formal grammar instruction. Most respondents favoured a non-directive role for CALL, with very few supporting the idea of the computer replacing the teacher (Levy 1997:127), but there was a significant lack of references to innovative pedagogical approaches:

Data-Driven Learning was the only new approach to language teaching that was cited by survey respondents as a direct result of the attributes of the computer. In other words, this approach has been conceived with the computer in mind. (Levy 1997:123)

There were, however, other types of programs and approaches that were directly attributable to the advent of the computer and already well established at the time of Levy's CALL Survey, for example: new ways of implementing action mazes, adventures and simulations (Higgins & Johns 1984:65–70), using the computer as an informant (Higgins & Johns 1984:70–74) and 'total Cloze' activities – which are currently assigned to the general category of text manipulation activities (Higgins & Johns 1984:155–177).

As for CALL methodology, the key to the successful implementation of CALL is summarised by Jones (1986) in an article with the significant title 'It's not so much the program: more what you do with it: the importance of methodology in CALL'. McCarthy (1999:4) describes the interrelationship between pedagogy and methodology as follows:

It is therefore not appropriate to make grand pronouncements or generalised value judgements about the pedagogical merits of a given piece of software. What is clear, however, is that software is not pedagogically inert. At one end of the CALL scale, tools such as word processing, concordancers, spell and grammar checkers and the Internet, do not exist in a vacuum. They are used to perform tasks within a program proposed by a teacher (with or without direct consultation with students), and that program, by design or default, reflects a pedagogical stance. At the other end come integrated learning systems, where the designer's methodology is an indissociable component of the courseware. Successful integration of CALL requires that some consideration be given to methodology.

The key to the successful use of any teaching materials in the classroom is *integration*. Assessing a new coursebook and planning how to use it is a relatively short process, as there are few variables to deal with. Integration of CALL materials into existing practice is more challenging than the integration of a new coursebook. It takes much longer for the teacher to assess the strengths and weaknesses of computer hardware and software and then to work out a classroom delivery strategy. As a first step, account has to be taken of:

- hardware currently available
- hardware that it would be desirable to have, e.g. multimedia labs for language teaching or interactive electronic whiteboards for whole-class teaching
- Internet access
- generic software currently available
- subject-specific software currently available
- software that it would be desirable to have
- classroom layout, e.g. rows of computers or 'islands' of 4–6 computers

At a theoretical level, account has to be taken of the teacher's knowledge about the different ways in which people learn languages. It is likely that the practising teacher has forgotten a good deal of what he/she learned about second language acquisition theories while undergoing initial teacher training and has developed his/her own set of ideas. A re-examination of existing theories and the teacher's own ideas in the light of the introduction of CALL materials into the classroom is therefore desirable.

Having considered the above, the teacher then has to consider how to set up CALL activities and how to integrate them into an existing scheme of work. It is likely that the CALL activities will constitute just a fraction of a week's work, so they have to be planned carefully. This means that, among other things, the teacher has to:

- decide at what point in the week CALL activities will be used
- decide whether to use CALL for whole-class teaching or to whether book a lab for individual or small-group work
- identify appropriate software packages and/or websites and check that they work
- author new materials if appropriate ready-made software packages or websites are not available
- print worksheets that may be used in conjunction with the CALL activities

The above are just a sample of the many things that need to be considered when integrating CALL into a program of study. Module 2.1 at the *ICT4LT Website* contains more detailed information, and Module 3.5 contains case studies of three secondary education and two higher education institutions that have successfully implemented CALL.

A teacher normally judges a language coursebook by the quality of its content, the soundness of its pedagogy, and the quality of its presentation. Few teachers would expect to be able to hand out a set of books to their students and ask them to get on with the work on their own. Most teachers would regard audio and video materials in much the same way as a coursebook, but there is a tendency to expect CALL materials 'to stand or fall on their own merits, without consideration of their role in a classroom lesson' (Jones 1986:171). CALL materials do offer considerable scope for independent learning (Little 2001 and Littlemore 2001), but this also requires careful planning. The important message that needs to be conveyed to teachers is that methodology is as important in CALL as it is in the use of any other kinds of teaching materials.

CALL authoring tools

Although authoring appears to be declining in popularity, easy-to-use template packages continue to be in wide use in schools and universities. Generic authoring packages such as Click2learn *Toolbook* and Macromedia *Director* have probably become too sophisticated, requiring an investment in time that language teachers simply do not have. Training in CALL authoring tools is offered on most ICT courses for language teachers, but is it really worthwhile?

Bickerton et al (2001) report on a survey of authoring tools, in which users were invited to evaluate a range of packages, considering functionality, ease of use, cost, and who was likely to use the package (e.g. programmer or classroom teacher). They conclude that more sophisticated authoring tools offering a wide range of functions and flexibility are more likely to require the user to engage in programming operations (Bickerton et al 2001:63). Such tools are best used by a team, one member of which already has programming skills. What most language teachers appear to require is a 'plug-and-play' authoring tool – a template that enables them to create simple exercises with the minimum of effort: see Bickerton et al (2001:60) for a list of examples. *Hot Potatoes* is another example of a popular set of authoring tools that can be described as 'teacher-friendly', offering the possibility of creating a range of simple exercises in a web environment. *Hot Potatoes* is dealt with in more detail by Arneil & Holmes (this volume).

Efforts are being made to simplify the authoring process for language teachers, for example in projects such as *MALTED* and the embryonic DISSEMINATE project (Delcloque 2001). The basic idea is that functionality and content are separated and that the teacher is empowered to assemble exercises or other material into pedagogically meaningful sequences or to create a whole course without having to engage in any programming operations. Access to an online assets base is a key feature of such projects. Piloting of *MALTED* in Spanish schools shows that teachers who are prepared to make the necessary investment in time can produce imaginative learning routines. For more detail see Bangs (this volume).

Teachers usually get an immense amount of satisfaction from using an authoring package that offers a pre-set range of templates for creating simple exercises. The satisfaction probably derives from a feeling that they have produced something of their own with relatively little effort. There is also a feeling of empowerment: the teacher provides the content and the computer does the work, namely delivering the exercises to the student and offering feedback. The main drawback of template packages is that they tend to be restrictive, reflecting to a large extent the pedagogical and methodological concepts of the person(s) who designed them. Tools such as *MALTED*, however, offer a wide range of interactivity – but, as already indicated above, a time factor needs to be taken into account in order to use such packages to their best advantage, for example to create multimedia activities or an integrated set of materials forming a complete course. Factoring feedback into the materials is especially time-consuming. For more detailed discussion of feedback issues see Bangs (this volume).

More creative teachers are likely to be dissatisfied with authoring tools that offer a pre-set range of templates and functionality, and they require more sophisticated authoring tools that enable them to control the authoring process. But once they begin to work with such tools they realise just how many factors they have to take into consideration in order to get acceptable results. The work produced by trainees who have attempted to use authoring packages such as *Toolbook* or *Director* tends to be strong on presentation, for example the use of colour, pictures and special effects, but weak in terms of pedagogy and interaction – underlining the need for including a module on pedagogy and methodology in an ICT training course for language teachers. This is not to say, however, that working with more sophisticated authoring tools is a waste of time. Teachers who have experienced what such tools offer and require are in a better position to work as part of a team, which might comprise a content provider (the teacher), a programmer, an instructional designer, a graphic designer and a sound/video technician.

A session on the use of authoring tools can be a valuable part of a training course for language teachers, but if time is short – and it usually is – then it is advisable to offer practical sessions dedicated to training in the use of authoring tools that offer a pre-set range of templates, followed by an introduction to more sophisticated authoring tools, emphasising the following key points:

- The teacher should be prepared to reject the use of the computer for a learning outcome best handled by other media, including pencil and paper.
- Pedagogy should always take precedence over technical issues.
- It needs to be decided whether sets of discrete exercises will be adequate, or whether a more sophisticated structure is needed, with conditional branching, remedial loops and so on.
- It needs to be considered how the materials created will fit into a wider curricular design.
- Copyright must be respected. Do not make unauthorised use of other people's assets.
- The time needed to author should never be underestimated.

Assessing trainees' work online

Thanks to email it is now relatively easy for trainees to submit work to their tutors online and for tutors to relay their comments back to their tutees. Tracking and reviewing facilities offered by *Word* enable the tutor to mark and comment on work submitted in DOC format, and tools such as *MARKIN* make the task even easier. Feedback from tutors involved in online training activities indicates that marking trainees' work is extremely time-consuming but an essential part of the training process.

Assignments submitted to tutors may take the form of a lesson plan, indicating:
- Aim of the lesson: e.g. to teach a specific point of grammar or vocabulary, or to engage learners in role-plays
- Hardware resources required: e.g. networked multimedia computer lab with web access or electronic whiteboard for whole-class teaching
- Software resources required: e.g. word processor, browser, a networked multimedia CALL package
- Preparation required: e.g. printed handouts, electronic worksheets, review of a website
- Learners' activities during the lesson
- Links to relevant websites
- Expected outcomes and follow-up activities

Examples of assignments that might be submitted under the four topic headings discussed above include:
- **Using generic software in language learning and teaching**
  (a) an electronic worksheet in *Word* format
  (b) a *PowerPoint* presentation for use with an electronic whiteboard
- **Exploiting the World Wide Web**
  (a) a description of a web quest
  (b) a description of vocabulary building activities using a web concordancer
- **CALL pedagogy and methodology**
  (a) an evaluation of a CALL software package or website
  (b) a discussion of the advantages and disadvantages of different pedagogical approaches and how ICT can be used to support them
- **CALL authoring tools**
  (a) a set of files produced with the aid of an authoring tool, including a discussion of the rationale behind them and how the materials would be integrated into a program of study
  (b) an evaluation and comparison of at least two different authoring tools

The above are based on actual assignments that have been submitted by trainees. Obviously, availability of local hardware and software restricts what trainees can be expected to do, and therefore assignments have to be negotiated with the tutor. Tutors need access to a wide range of tools in order to be able to assess trainees' work, and experience shows that they also need to be well protected against viruses. Trainees, particularly those working from home, are somewhat lax at updating their anti-virus software. A good deal of awareness-raising about web security and netiquette is an essential part of the training process.

Some solutions to training issues

This chapter concludes with an overview of four important contributions made to the understanding and designing of ICT training materials for language teachers, including feedback on successes and failures.

1. The New Opportunities Fund (NOF) training initiative

One of the most extensive projects ever undertaken in the provision of in-service ICT training for teachers is the *New Opportunities Fund (NOF)* initiative in the UK. The NOF initiative was allocated 230 million pounds (around 370 million euros) of National Lottery money, with a nominal sum of 450 pounds (around 720 euros) available for training each full-time teacher in the state primary and secondary school sectors across all subject areas. The NOF training program has been in operation since April 1999 and will come to an end in December 2003. Online training has featured in most of the individual programs offered by approved training providers.

The NOF training initiative followed a substantial government program of investment in hardware and online services, including:

- Providing hardware for schools. By 2001 the student to computer ratio in secondary schools was 8:1, and in primary schools the ratio was 13:1. The respective targets for August 2002 are 7:1 and 11:1.
- Connecting schools to the Internet. By 2001 96% of primary schools and all secondary schools were connected to the Internet.
- Setting up the *National Grid for Learning (NGfL)*, a large online resource bank of information for teachers.
- Purchasing laptop computers for teachers.
- Setting up the *Curriculum Online* project, which involves the production of digital learning resources across all subjects.

The NOF training program has been subjected to close scrutiny by the Office for Standards in Education (OFSTED), which has produced two progress reports since the initiative began. The most alarming finding contained in the second progress report is that, in spite of the substantial investment in NOF training, it has only had a significant impact in a quarter of secondary schools and one third of primary schools (OFSTED 2002:22). These are overall figures; figures for specific subject areas are not available. Training *can* be effective if the conditions are right, for example:

- Training is most successful where senior managers in schools take an active interest in teachers' progress, where there is effective peer support, and where groups of teachers meet for part of the training. (OFSTED 2002:3)
- Successful online mentoring operated at ratios of under 30 trainees to each mentor. (OFSTED 2002:23)
- Personal individual access to a computer by teachers, especially at home, has continued to be one of the strongest influences on the success of ICT training. (OFSTED 2002:24)

Several reasons are cited for the failure of training to make a significant impact (OFSTED 22–26), for example:

- Motivation has waned in many teachers where they have not obtained appropriate subject advice and guidance.
- Many teachers have found online support to be unsatisfactory. This was usually because access was unreliable or because mentors were dealing with too many teachers and their responses were therefore often infrequent, shallow or unhelpful.
- The reasons why teachers and schools fail to persevere with training vary widely. The most frequently cited by teachers include lack of time, technical and organisational difficulties, poor support from trainers or mentors, poor match of training materials to needs, expectations to complete exercises or compile portfolios that are unrelated to teachers' current work, lack of good subject-specific ideas and resources.
- Teachers who were left to their own devices to use distance learning materials on CD-ROM frequently made little headway and did not complete the training.

These are valuable lessons here for ICT training initiatives worldwide. Whether those responsible for funding and delivering training will take any notice of them remains to be seen.

2. CILT-NOF

One of the reasons for dissatisfaction with the NOF training program commonly cited by teachers is that training providers failed to take proper account of individual subject teachers' needs. Most NOF training providers tried to cover too many different subject areas. Those providers who concentrated on one subject area have been more successful, for example the Centre for Information on Language Teaching (CILT), London, which offers an accredited *CILT-NOF* introductory course for language teachers consisting of:

- two contact days of face-to-face training – one day at the start of the course and the second day at the end;
- eight weeks of one-to-one distance learning with an online tutor, including access to a bank of web-based materials containing extensive subject-specific reading and tasks for completion.

The topics covered by the *CILT-NOF* course are:

- Using word processors in language learning and teaching
- Text manipulation
- Using databases in language learning and teaching
- Whole-class teaching using *PowerPoint* and interactive whiteboards
- Email
- The World Wide Web
- Evaluating ICT resources

Information provided by CILT indicates that this model, namely the combination of face-to-face and online tuition, has proved extremely successful. The drop-out rate has been less than 10%, which paints a very different picture from that emerging from reports on the NOF initiative as a whole. Some teachers report on the benefits gained from the opportunity to reflect on their general language teaching practice and the potential learning gains through ICT. Others report on a gain in confidence in dealing with the technical support staff in their schools and in claiming equal access to ICT facilities vis-à-vis other subjects. As for peer-group support, many teachers did not have the time to contribute to a forum for trainees in addition to the email contact they were having with their tutor, and for some teachers the pressure of work in their schools meant that a distance-learning course was untenable during term time. It was felt that more flexible deadlines should be offered, and that the eight-week time limit should be extended.

Feedback from individual tutors on the course indicates that the following improvements would be welcomed:
- An effective system for assessing teachers' basic and applied subject-specific skills before they embark upon an online course is essential.
- A standardised system of assessing trainee's competence at the end of an online course is essential.

3. The ICT for Language Teachers (ICT4LT) project

The ICT4LT project was initiated in early 1998 by a consortium of four European universities and the Centre for Information on Language Teaching (CILT) in London, coordinated by Thames Valley University and assisted by funding provided under the Socrates Program of the European Commission. The aim of the project was to provide a set of online training materials in ICT for language teachers in three different countries. The outcome is the *ICT4LT Website* and a set of printed materials in four different languages (English, Italian, Finnish and Swedish), plus an offline version of the website on CD-ROM (in English only). The training modules that have been created at the website have been designed for a wide range of trainees with differing levels of experience. Table 4 shows the current set of English-language modules (August 2002):

Table 4. ICT4LT modules

Basic Level

Module 1.1: Introduction to new technologies

Module 1.2: Introduction to computer hardware and software

Module 1.3: Using text tools in the MFL classroom

Module 1.4: Introduction to CALL

Module 1.5: Introduction to the Internet

Intermediate Level

Module 2.1: CALL methodology: integrating CALL into study programs

Module 2.2: Introduction to multimedia CALL

Module 2.3: Exploiting WWW resources online and offline

Module 2.4: Using concordance programs in the MFL classroom

Module 2.5: Introduction to CALL authoring programs

Advanced Level

Module 3.1: Managing a multimedia language centre

Module 3.2: CALL software design and implementation

Module 3.3: Creating a WWW site

Module 3.4: Corpus linguistics

Module 3.5: Human Language Technologies

Additional Module

Module 4.1: Computer aided assessment (CAA) and language learning

A regular analysis of accesses to the modules is carried out in order to assess visitors' needs and to help determine the content of updates to the website. This analysis is published on the English-language homepage of the *ICT4LT Website*. Recent analyses reveal some interesting trends:

- A high proportion of visitors to the website appear to be novices. The Basic Level modules – apart from the introductory module on hardware and software – are the most visited modules. Records indicate that these modules are used by a number of universities and teacher training colleges as the basis of introductory courses in ICT and language learning and teaching.
- The module on multimedia CALL has been the most visited module since it first appeared.
- The module on managing a multimedia language centre has remained consistently in the top five most visited modules.
- There is an increasing interest in the module dealing with the use of concordance programs in the languages classroom.
- There is an increasing interest in the module on computer-aided assessment.
- There is a declining interest in the modules dealing with authoring programs and CALL software design and implementation.
- The module on CALL methodology has consistently been the least visited module since it first appeared.

When the ICT4LT project was initiated it was intended that it would lead to a recognised qualification or set of qualifications. It was assumed that an assessment scheme would be put in place and that online tutors would be appointed. As the project developed, however, it was realised that the funding awarded by the European Commission to set up the project would not cover online tutoring. Paying regular online tutors for their services proved to be much more expensive than anticipated, and it seems unlikely that the site will generate enough income to cover their costs, so for the moment the website consists of a set of free materials with occasional tutor support. There is, however, a Bulletin Board at the website and the *ICT4LT JISCMail Discussion List* is also available to site visitors. Learning Tasks and Discussion Topics are inserted at appropriate points in each module, and visitors to the site are invited to submit solutions to the Learning Tasks and to express their views on the Discussion Topics. Questions addressed to the project management team are quickly answered. However, very few questions are actually put to the management team – no more than one or two per month, in spite of the site's access rate of around 400–500 visitors per day, and the Discussion List is virtually dead. The Bulletin Board springs to life on occasions, usually in conjunction with face-to-face courses that have just taken place. Regular tutor online support is clearly needed in order to sustain interaction.

In the longer term it is anticipated that the ICT4LT materials will form the backbone of an accredited online course. In the meantime the materials are proving to be invaluable as a support to one-day face-to-face training courses that are delivered on an ad hoc basis:

- as study materials *before* trainees participate in face-to-face training.
- as materials on which trainees work *during* face-to-face training.
- as follow-up study materials *after* trainees have completed face-to face training.

This is similar to the model adopted by the *CILT-NOF* course, but it has not been as widely piloted. Feedback, however, has proved positive.

4. TALLENT

The TALLENT project (Teaching and Learning Languages Enhanced by New Technologies) was initiated in early 1998 by a consortium of eleven European universities, coordinated by the University of Limerick and assisted by funding provided under the Socrates Program of the European Commission. The aim was to design a syllabus in ICT for language teachers that could be delivered via workshops and seminars at universities in different European countries. The current syllabus can be consulted at the *TALLENT Website*. A printed handbook is also envisaged.

The course is not intended to be delivered online; it is a residential course and delivered face-to-face over five or ten days, with 30 class-contact hours per five-day period. The course consists of three core pedagogical elements that should be included in all TALLENT courses, with additional optional elements. The course offers a negotiated syllabus, following an analysis of the needs and interests of each target group:

Core elements of the course:

- Introduction and needs analysis – participants are invited to complete a questionnaire before attending the course to assist tutors to adapt materials and teaching to their needs
- Language learning and ICT: theoretical background, pedagogy and methodology
- The self-directed learning environment and ICT

Optional elements of the course:

- Reference tools: online dictionaries, databases and library resources
- Using the Internet in language learning and teaching
- Creating a web page for language learning
- Authoring tools: evaluating, selecting and using them
- Concordancing: data-driven learning, using and creating corpora
- Speech in action: using ICT to teach pronunciation and listening
- Setting up and running a language resource centre

Littlemore (2002) reports on a TALLENT course delivered at the University of Birmingham in 2001, analysing participants' reactions and identifying strengths and weaknesses of the course. Creating a web page for language learning was the most popular module, following by the module dealing with authoring tools. The least popular module was that concerned with setting up a language resource centre. Some aspects of the course were criticised because of problems associated with running multimedia applications rather than the content itself. This appears to be a widespread problem; many trainers and trainees have reported on the failure of university and school computer services departments to set up their computer labs to handle sound/ video output and sound input adequately. Theoretical aspects of the TALLENT course, dealing with pedagogy and methodology were not rated highly, but this may have been a question of presentation. Among Littlemore's many recommendations for future courses, the following stands out:

> *Instructors should present real examples from their own teaching experience; programs should be introduced in a 'hands-on', step-by-step manner, allowing participants to create their own materials as they go along; and participants should complete each module with a piece of teaching material that they can use with their students. (Littlemore 2002:7)*

Conclusion

There is no question that high-quality training is an essential element in the process of introducing ICT into language learning and teaching. It is also an ongoing process, requiring regular updates. Online training is playing an increasingly important role, but practical aspects of training can be delivered better in face-to face workshops. A judicial mix of online and face-to-face training is therefore desirable. Online training works best when there is substantial peer group and tutor support. The technology for delivering online training must be robust, the user interface must be transparent, and hardware must be easily accessible to trainees. Content must be relevant and consist of a mix of theory and practical aspects. Trainees need adequate time to complete assignments set by tutors, and tutors need time to mark them. Distance training is a labour-intensive delivery medium, and this has to be carefully costed. Above all the needs of the students have to be borne in mind when setting up an online course; it is for their benefit, not for the benefit of educational administrators. Training is not cheap, but it is more expensive in the long term not to invest in training.

References

Bickerton, D., Stenton, T. & Temmermann, M. (2001). Criteria for the Evaluation of Authoring Tools in Language Education. In A. Chambers & G. Davies (Eds.), *Information and Communications Technologies in Language Learning: A European Perspective*. Lisse: Swets & Zeitlinger.

Davies, G.D. (1997). Lessons from the Past, Lessons for the Future: 20 Years of CALL. In A-K. Korsvold & B. Rüschoff (Eds.), *New Technologies in Language Learning and Teaching*. Strasbourg: Council of Europe. Also available (2002) in a revised edition at: http://www.camsoftpartners.co.uk/coegdd1.htm

Davies, G.D. (2002). ICT and Modern Foreign Languages: Learning Opportunities and Training Needs. *International Journal of English Studies,* 2 (1). Monograph Issue, *New Trends in Computer Assisted Language Learning and Teaching.* Murcia, Spain: Servicio de Publicaciones Universidad de Murcia. http://www.camsoftpartners.co.uk/needs.htm

Delcloque, P. (2001). To DISSEMINATE or not? Should we Pursue a New Direction: Looking for the 'Third Way' in CALL Development? In A. Chambers & G. Davies (Eds.), *Information and Communications Technologies in Language Learning: A European Perspective.* Lisse: Swets & Zeitlinger. See also the related website: http://www.disseminate.org.uk

Felix, U. (2001). *Beyond Babel: Language Learning Online.* Melbourne: Language Australia.

Gitsaki, C. & Taylor, R. (1999). *Internet English: WWW-based Communication Activities.* Oxford: Oxford University Press.

Higgins, J. & Johns, T. (1984). *Computers in Language Learning.* London: Collins.

Jones, C. (1986). It's not so Much the Program: More What You Do with It: The Importance of Methodology in CALL. *System,* 14 (2), 171-178.

Kirkman, G., Osorio, C. & Sachs, J. (2002). The Networked Readiness Index (NRI): Measuring the Preparedness of Nations for the Networked World. In G. Kirkman, J. Sachs, K. Schwab & P. Cornelius (Eds.), *Global Information Technology Report 2001-2002: Readiness for the Networked World.* Oxford: Oxford University Press. http://www.cid.harvard.edu/cr/gitrr_030202.html

Levy, M. (1997). *Computer Assisted Language Learning: Context and Conceptualisation.* Oxford: Oxford University Press.

Little, D. (2001). Learner autonomy and the challenge of tandem language learning via the Internet. In A. Chambers & G. Davies (Eds.), *Information and Communications Technologies in Language Learning: A European Perspective.* Lisse: Swets & Zeitlinger.

Littlemore, J. (2001). Learner Autonomy, Self-instruction and New Technologies in Language Learning: Current Theory and Practice in Higher Education in Europe. In A. Chambers & G. Davies (Eds.), *Information and Communications Technologies in Language Learning: A European Perspective.* Lisse: Swets & Zeitlinger.

Littlemore, J. (2002). Setting Up a Course in ICT for Language Teachers: Some Essential Considerations. *CALL-EJ Online,* 4 (1). http://www.clec.ritsumei.ac.jp/english/callejonline/7-1/littlemore.html

McCarthy, B. (1999). Integration: The sine qua non of CALL. *CALL-EJ Online,* 1 (2). http://www.clec.ritsumei.ac.jp/english/callejonline/4-2/mccarthy.html. Reproduced (2002) with the author's permission at http://www.ict4lt.org/en/McCarthy.htm

OFSTED (Office for Standards in Education) (2002). *ICT in Schools: Effect of Government Initiatives: Progress Report.* http://www.ofsted.gov.uk

Oppenheimer, T. (1997). The Computer Delusion. *The Atlantic Monthly,* 280 (1), 45–62. http://www.theatlantic.com/issues/97jul/computer.htm

Vogel, T. (2001). Learning out of Control: Some Thoughts on the World Wide Web in Learning and Teaching Foreign Languages. In A. Chambers & G. Davies (Eds.), *Information and Communications Technologies in Language Learning: A European Perspective.* Lisse: Swets & Zeitlinger.

Warschauer, M. (1996). Computer Assisted Language Learning: An Introduction. In
  S. Fotos (Ed.), *Multimedia Language Teaching*. Tokyo: Logos International. http:/
  /www.gse.uci.edu/markw/call.html
Windeatt, S., Hardisty, D. & Eastment, D. (2000). *The Internet*. Oxford: Oxford
  University Press. See also the related website: http://www1.oup.co.uk/elt/
  rbt.Internet

Websites

BBC Languages – http://www.bbc.co.uk/education/languages
Bonjour – http://www.bonjour.org.uk
CILT-NOF – http://www.cilt.org.uk/nofict/index.htm
Curriculum Online – http://www.curriculumonline.gov.uk
Dániel Berzsenyi College – http://www.bdtf.hu
EUROCALL – http://www.eurocall.org
Hot Potatoes – http://web.uvic.ca/hrd/halfbaked
ICT4LT Website – http://www.ict4lt.org
ICT4LT JISCmail Discussion List – http://www.jiscmail.ac.uk/lists/ict4lt-pilot.html
MALTED – http://www.malted.com – http://malted.cnice.mecd.es
MARKIN – http://www.cict.co.uk/software/markin
National Grid for Learning (NGfL) – http://www.ngfl.gov.uk
New Opportunities Fund (NOF) – http://www.nof.org.uk
TALLENT Website – http://www.solki.jyu.fi/tallent
Web Literacy – http://www.ecml.at/projects/voll/literacy
Web Skills for Language Learners – http://www.well.ac.uk/wellproj/booklet/
    booklet.htm
WorldCALL – http://www.worldcall.org

All websites in this chapter were verified on 20.09.2002.

Index

A

Achill Project   125
Adobe's Atmosphere   127
anarchitexture   116
audiographics   150, 160
authentic assessment   13, 159, 160
authoring prototype   28
authoring tools   21, 23, 25, 36, 64-66, 69, 71, 75, 76, 89-92, 95, 196, 204-206, 211, 212
automated feedback   151, 159
Automatic Speech Recognition (ASR)   201
avatar-centred and place-centred environments   126

B

Blackboard   10, 44-46, 50, 52-55, 59, 61, 66
Bloom's taxonomy   133, 134

C

Centre for Information on Language Teaching (CILT)   208, 209
client-side exercises   11, 64-66, 68
client-side interactivity   52
client-side self-access learning exercises   59, 60
cognitive walkthrough   27
collaborative concept map module   178
collaborative whiteboard   177
communicative model   149
complementary training   194
comprehensible input   48, 162, 172
comprehensible output   172
concept maps   184
conceptual model   22, 24-27, 32-34
conditional branching   11, 88, 92, 205
constructivist learning   15, 83, 123, 137, 138, 165, 173
constructivist projects   157
Co-operative Learnware Object Exchange (CLOE)   70
course shell   45

cultural simulation: Virtual weddings and a real wedding of linguistics, literature and cultural studies   129

Curriculum Online project   207

CU-SeeMe   175

D

design cycle   24, 92

design decision template   35

design space   21, 25

DHTML   52

discovery learning   60, 68

discussion forums   44, 45, 51, 52, 171

DISSEMINATE project   204

Dublin Core standard   70

E

elaborative feedback   82

European Commission's TEMPUS/PHARE/TACIS program   195

EVE (English Virtual Environment)   124, 131-133, 140

extrinsic and intrinsic feedback   11, 82

F

feedback   1, 9, 11, 13-15, 23, 24, 30, 36-38, 40, 52, 53, 60, 62-65, 68, 71, 72, 81-83 84-95, 108, 110, 115, 123, 125, 148-153, 159, 165, 196, 200, 201, 205-207, 209, 211

frames environment   45, 46, 48, 49, 50

functional model   27-29

G

graphical virtual environments   12, 124, 126, 137

H

Hexe Hilde   155, 157, 162

hierarchical task analysis diagram   32

Hot Potatoes   11, 14, 52, 53, 59, 60, 64-66, 69, 71, 74, 91, 92, 95, 157, 204

HTML   44, 49, 52, 59, 69, 72, 90-93, 179, 188

I

ICT for Language Teachers (ICT4LT) project   209
IEEE Learning Technology Standards Committee   72, 73
IMS Consortium   69, 72
IMS Global Learning Consortium   71
IMS standard   71, 74
individualised feedback   151
information architecture   22
instructional designs   90
Intelligent CALL (ICALL)   153
Intelligent Language Tutoring System (ILTS)   152
interaction design   22, 23, 29

K

Krashen's input hypothesis   172

L

language anxiety   13, 160, 164
learner control   157, 158
learner tracking   70, 71, 73
learning environment   7, 9, 11, 21, 25, 50, 55, 174, 183, 188, 200, 211
Learning Management System (LMS)   10, 44, 45, 50, 53, 54, 59, 61, 76, 94
learning styles   154, 157, 164, 174
Levi-Strauss's concept of the bricoleur   123
LGU authoring tool   28, 35
Lip-synching technology   127

M

MALTED (Multimedia Authoring for Language Tutors and Educational Development)
    11, 14, 93-95, 204, 205
MALTED authoring tool   81
mental models   22, 27-30, 38
metacognitive reflection   152
metadata   69-72, 76, 77, 94
metalinguistic feedback   152
Microsoft NetMeeting   182
MOO-based learning   97, 98, 107, 114, 116
MOO-based tandem projects   107
MOOmail   105, 106, 109, 112, 113, 116
Multi User Domain – Object Oriented   97

N

National Grid for Learning (NGfL)   207
Natural Language Processing (NLP)   152
negotiation of meaning   172
netLearn environment   46
NetLearn Languages (NLL)   182, 186, 188
NetMeeting   175-178, 180, 182, 184
Networked Readiness Index (NRI)   194
New Opportunities Fund (NOF)   196, 207
Norman's framework   24

O

object oriented   97, 98, 105, 125
online communities   157
online testing   67

P

packaging   69, 71, 72
peer assessment   159, 160
performance tracking   65
problem space   25, 26, 33
problem-based learning activities   86
process-oriented learning   148

Q

QTI Lite   71, 72
quality indicators for online learning   150
Question & Test Interoperability specification   71
Quia   157

R

ReLaTe (Remote Language Teaching)   175, 176, 182, 183

S

SCORM (the Sharable Content Object Reference Model proposed by the Advanced
    Distributed Learning organisation)   72, 75-77
sequencing   36, 70-72
Server-hosted database-driven exercises   59, 61
server-side exercises   11, 59, 68
site organization and navigation   45

situated learning   148, 173
social dilemma   13, 157, 158
speech tag parser   154
structural model   27, 28
Structure Chart   32
Student–teacher constellations   182
summative evaluation   37, 38
Swedish community platform Lunarstorm   123
synchronous online communication and collaboration   174

T

TALLENT (Teaching and Learning Languages Enhanced by New Technologies)   196, 200, 211, 212
tandem MOO   108, 109
telnet MOO client   98
template-based web course authoring and management system   44
Text-based MUDs (Multi-User Dungeons)   125
think aloud protocols   27
transactional distance   158
twin classification   27

U

usability feedback   24
usability goals   23, 25, 31, 33, 34, 40
user acceptability   21, 40
user-centred   10, 21

V

verification feedback   82
virtual village   111
Virtual Wedding Project (VW Project)   7, 123, 124, 128, 129, 132-134, 138, 149, 142
virtual world   11, 14, 113, 123-129, 131, 134-137, 140, 149, 157, 164
virtuality   126
Visiting EVE   131
voice recognition software   160
voiced bulletin boards   160
voiced chats   149, 160
Vygotsky's sociocultural theory   173

W

web archives 49

web links **44**, 45, 48

Web Literacy tutorial materials 199

Web Skills for Language Learners tutorial materials 199

web-based MOO client 98

WebCT 10, 44-46, 49, 50, 52-55, 59, 61, 64, 66, 69, 93

WorldCALL 195

X

XML **69**, 70-72, 74, 76, 77, 90, 93

XSLT 77

XSLT stylesheet 74

LANGUAGE LEARNING AND LANGUAGE TECHNOLOGY
ISSN 1568-248X

1. ICT and Language Learning: A European Perspective
   Edited by Angela Davies and Graham Chambers
       2001   ISBN 90 265 1809 9 (hardback)
                 ISBN 90 265 1810 2 (paperback)

2. The Changing Face of CALL: A Japanese Perspective
   Edited by Paul Lewis
       2002   ISBN 90 265 1934 6 (hardback)
                 ISBN 90 265 1935 4 (paperback)

3. Language Learning Online: Towards Best Practice
   Edited by Uschi Felix
       2003   ISBN 90 265 1948 6 (hardback)